TIGER TANK

© The Tank Museum 2011 and 2020
Text by David Fletcher and David Willey
Additional text by Mike Hayton, Mike Gibb, Darren Hayton, Stevan Vase
and David Schofield

All rights reserved. No part of this publication may be reproduced or
stored in a retrieval system or transmitted, in any form or by any means,
electronic, mechanical, photocopying, recording or otherwise, without
prior permission in writing from the Publisher.

Previously published in hardback in a larger format in 2011 as the
Tiger Tank Owners' Workshop Manual
This edition published in 2020
Reprinted May 2022 and April 2024

David Fletcher, David Willey, Mike Hayton, Stevan Vase, Mike Gibb,
Darren Hayton and David Schofield have asserted their moral rights to be
identified as the authors of this work.

A catalogue record for this book is available from the British Library

ISBN 978 178521 687 9

Library of Congress control no. 2019945933

Published by Haynes Group Limited,
Sparkford, Yeovil, Somerset BA22 7JJ, UK
Tel: 01963 440635
Int. tel: +44 1963 440635
Website: www.haynes.com

Haynes North America Inc.,
2801 Townsgate Road, Suite 340
Thousand Oaks, CA 91361

Printed in India.

TIGER TANK

David Fletcher, David Willey, Mike Hayton, Mike Gibb,
Darren Hayton, Stevan Vase and David Schofield

HAYNES ICONS

THE TANK MUSEUM

Contents

Foreword

My first encounter with the German Tiger I was on 21 April, 1943, when, as a young commander of a Churchill tank troop in A Squadron, 48th Battalion Royal Tank Regiment, I was one of two troop leaders leading a counter-attack against a German defensive position in two hills near Medjez-el-Bab, Tunisia. This was the Battalion's first action since landing in North Africa and was part of the First and Eighth Armies' final push to clear Tunisia of Axis troops and to take Tunis. We had, of course, heard mention of the Tiger, but very little had penetrated down the food chain to troop leader level; about the dreaded '88' gun, however, we had not only heard a lot but

had also seen its effect on the Churchill tank when we arrived at the railhead and saw the stacked hulks of the tanks from 25 Army Tank Brigade with large holes punched through even their thickest armour.

As we advanced towards our objectives we could see no sign of enemy, but suddenly my fellow troop leader's tank erupted in an enormous explosion, blowing the turret crew out of the turret and setting the tank on fire. Before I had time to locate the source of this shot my own tank was hit by an 8.8cm shot which passed through the front plate right down the length of the tank and into the engine, setting it on fire. I and my crew were luckier than my colleague's as we all managed to bail out with only minor injuries. The following day we drew a replacement tank and took the opportunity of examining our burnt-out one; it had been cleanly penetrated by an 8.8cm armour-piercing shell, which we later discovered, when examining the battlefield, had been fired by a Tiger tank that was found abandoned and virtually undamaged on our objective. It had been hit by a 6pdr shot from one of our tanks that had ricocheted into the turret ring preventing the gun from traversing.

The capture of a running Tiger in virtually undamaged condition was a prize of great value, which would give our own tank designers much valuable information when shipped back to the UK for detailed examination. It was moved to Tunis where it was refurbished from captured German spares and visited by HM King George VI, the Prime Minister Winston Churchill and many other VIPs before being shipped back to the UK and the School of Tank Technology at high priority for detailed examination and reporting.

I meanwhile had also been shipped back to the UK as a result of injury sustained when my tank was hit and on arrival was posted as a student to the School of Tank Technology at Chobham to attend the next Advanced

Class in Tank Technology, due to assemble in March 1944. Pending the commencement of the course, I was taken on the staff of the School as an examiner of captured foreign AFVs, specialising in armament, stowage and fighting arrangements. Imagine my amazement when the Tiger that had caused me so much trouble in North Africa turned up at the School for examination in October 1943; it was an extraordinary coincidence, which resulted in my writing the examination report on the very gun that had knocked me out. My examination report on its armament, stowage and fighting arrangements was published in the final report in January 1944, but I felt no anger towards

it; in fact I became quite fond of it, and when the Tiger was taken over by the Tank Museum I took an interest in its future as, at that time, as senior instructor at STT after the war I had a supervisory rôle in the Tank Museum's management. I tried to get the tank refurbished, but at that time in the 1960s there was no money available, but I am delighted now to see it in pristine condition once again, thanks to the National Lottery, and have since been involved in several TV programmes with it. I therefore recommend the publication of this manual, describing what is now my favourite tank.

Peter Gudgin
January 2011

BELOW Face to face with a Tiger: Peter Gudgin meets Tiger 131 again, the tank that almost killed him in 1943. *(BBC)*

Introduction

For a tank that was used in relatively small numbers during the Second World War, an awful lot has already been written about the German Tiger. However, the Tank Museum at Bovington is in a unique position to write a new chapter in the Tiger story with its own discoveries on restoring and running Tiger 131, which forms the centrepiece of this manual.

OPPOSITE Even on a sunny day in Dorset, the Tiger has an air of potent menace as it motors towards you. *(All photographs are copyright © The Tank Museum unless credited otherwise)*

restoring and running Tiger 131. We are also in a unique position to fill the book with images of the restoration in this very successful Haynes format.

To write anything on the Tiger in English, one inevitably refers to certain authors and publications. We have, of course, used a number of the wartime official manuals issued for the tank including the now famous *Tigerfibel* (German Tiger Tank manual).

The original reports made during the war by the School of Tank Design still make wonderfully clear, concise reading and are beautifully illustrated. Peter Gudgin, who has written the foreword to this book, was at the engagement in Tunisia in 1943 when Tiger 131 was captured. With remarkable coincidence, after convalescence he was tasked with writing major parts of the British report on the tank. These reports have subsequently been published with David Fletcher's commentary and David, with his huge wealth of knowledge and experience in writing on armour, has written the first chapter in this book.

Later writers have given us much more; over the last thirty years the partnership of Tom Jentz and Hilary Doyle, researching and drawing

Another book on the Tiger? For a tank that was used in relatively small numbers during the Second World War an awful lot has been written about the Tiger. At the Tank Museum our shop reflects that interest, selling publication after publication on the Tiger and other German vehicles; Allied tanks come a poor second. Our excuse for this book is to hopefully add another angle to the Tiger story with the inclusion of our own discoveries on

German vehicles from original sources, has been outstanding. Tom's insistence in only using contemporary source material has provided valuable evidence and factual information that helps challenge many of the myths that inevitably grow around a subject area over time. He has been of great help in the museum's and others' attempts to gradually return German vehicles to more original condition. His presentation of the material on the Tiger and his statistical information has been drawn upon heavily in this book. If readers want to study more on the tank, I can only suggest they seek out Tom and Hilary's publications.

Our thanks go these authors on the subject and for any errors and myths that we add to or perpetrate in this book, my apologies.

And remember …

In the Tank Museum collection there is a set of campaign medals awarded to 7091246 Lance Corporal William Francis Aspinall. In 2007 the museum bought the set of four medals from eBay for just over £200.

Sadly little is known of Aspinall, but we do know he joined the 48th Royal Tank Regiment (48 RTR) and in consequence was at Djebel Djaffa on the afternoon of 21 April 1943. As the gunner of Captain Alan Lott's Churchill, he was part of the A Squadron attack. The Churchill hit a Panzer III then in turn was hit by a round that penetrated just below the gun mantlet, causing an explosion. Aspinall, the driver Trooper Bernard Marriott, and the co-driver, Trooper Richard Smith, were all killed. Alan Lott was badly burnt and died two weeks later in hospital. Aspinall was 24.

Aspinall and his colleagues who died that day fought for an ultimate cause that even in today's non-judgemental era shines out as a truly worthy fight. He was on the right side; the others, however brave, were not.

While we can become fascinated by the design, the ingenuity and sheer presence of a major German vehicle like the Tiger, we cannot let ourselves forget what the machine was made for and the regime that used it. To do so would be of great disservice not just to the memory of the likes of William Aspinall, but it also diminishes our own intelligence.

David Willey, Curator, Tank Museum
February 2011

ABOVE It's a bloke thing. There is just something about tanks that dads and sons can relate to.

BELOW The Curator, David Willey, in younger days. *(David Willey)*

'Tiger hulls were assembled at the Henschel factory in Kassel, where railway locomotive production was compelled to share factory space with the Tiger – locomotives being made on the left side, Tigers on the right. On average, there were always 18 tanks in the hull assembly workshop, and 10 on the final assembly line.'

David Willey, Tank Museum Curator

Chapter One

The Tiger Story

David Fletcher

It is tempting to view the appearance of the Tiger tank as revolutionary, as indeed it must have seemed to the many Allied soldiers who had to face it. Looked at from the standpoint of history, and in particular engineering history, it was purely evolutionary – not always at the cutting edge of development, sometimes reactionary, but the end result was a machine that represented a quantum leap forward in tank design and changed everything – forever.

OPPOSITE The Tank Museum's Tiger 131 in Tunisia soon after capture, minus its front left road wheel. Front road wheels were sometimes removed by crews to lessen the risk of mud and stones jamming between the wheels, causing wheel damage or tracks riding up and off the front sprocket wheel. Battle damage can be clearly seen on the *Feifel* air cleaners at the rear. A 48 RTR Churchill and Daimler Dingo scout car can be seen in the background.

RIGHT The French Char FCM 2C had its origins in the First World War. Only ten were built but they attracted German attention during the interwar period.

The pioneers

The Tiger evolved from a design that was first discussed in October 1935 for a 30-tonne tank armed with a high-velocity 75mm gun, capable of defeating the massive French Char 2C and its derivatives. It seems an odd choice to regard as a potential threat in 1935. The 2C had been designed towards the end of the First World War and only ten had been completed before production ceased with the return of peace. True it was a monster, over 10m long, 4m high and weighing 69 tonnes, with guns poking out all over the place. Even so maximum armour thickness was only 45mm and in terms of tactical mobility the Char 2C was more a liability than an asset. Presumably as far as the Germans were concerned it represented the greatest threat imaginable at that time. Whether this had anything at all to do with Hitler's ultimate rise to power a year earlier is a moot point.

No actual tank, not even a prototype, resulted from these early deliberations although two points arise from the discussions, both

relating to the engine, which might be worth mentioning; firstly it should be noted that, even at this early stage, interest focused entirely upon petrol (gasoline) engines. Diesel power units are never discussed. The reason seems to be that the German Army required a certain standard of performance over a given period of time which only a petrol engine could deliver. Yet the idea that German tanks were powered by diesel engines persists. The other concern involved the size of engine and the power it could deliver. What the authorities wanted was a six-cylinder engine with a potential maximum delivery of 700hp which was comfortable under normal circumstances at 600hp. Their favoured consultant was the Maybach Company governed by the legendary Wilhelm and his gifted son Karl. They regarded the attainment of these figures, from a six-cylinder engine, as fantastic, without a long development programme and suggested that the lower figure might be achieved with a V-12 and the higher figure only possible with a V-16. This, however, involved a penalty that has bedevilled tank designers ever since. A V-16, for example, would require a much larger engine bay, which in turn meant a longer tank, so that weight would increase to the point that any extra horsepower delivered by the bigger engine would be absorbed by the greater weight. As things stood it was a pointless exercise.

The sequence of events that led, ultimately, to the appearance of the Tiger is both complicated and convoluted. Typically various firms were drawn into the programme and, in order to explore different ideas, were to some extent pitted against one another, resulting in a

VK 20.01 (D).

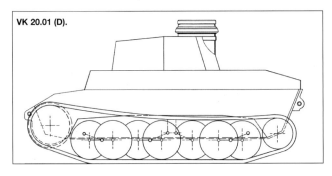

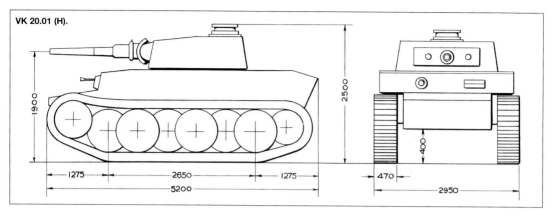

VK 20.01 (H).

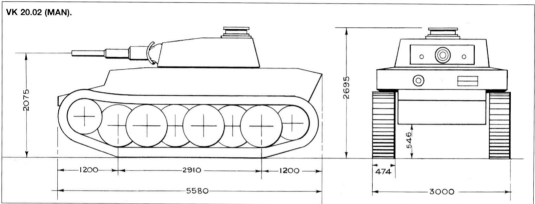

VK 20.02 (MAN).

number of prototypes, many of which rapidly became redundant. The process began with a trial chassis from Henschel designated D.W. 1 (*Durchbruchswagen*) or breakthrough tank for the direct assault role, to be fitted with a trial turret, designed by Krupp, mounting a short 75mm gun.

The D.W. series was later reclassified VK 30.01, indicating a tank in the 30-tonne class and the outward appearance was not dissimilar to the Panzer IV except in the matter of suspension. Although many alternatives were investigated the Germans were clearly most impressed by the characteristics of torsion bars, running across the floor of the hull, and it proved possible, by placing these fairly close together, to get more wheels per side, thus evening out pressure points on the tracks and spreading the weight. This was only achieved by overlapping and interleaving alternate sets of wheels in a fashion that became characteristically and almost uniquely German.

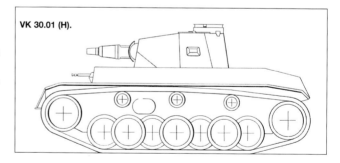

VK 30.01 (H).

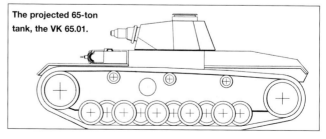

The projected 65-ton tank, the VK 65.01.

RIGHT VK 30.01 (H).

Meanwhile, in both Germany and Britain designers were at work on ever more complex projects for tank steering systems. In Germany a number of light tank projects, starting in about 1937, included multiple gear steering systems (presumably in the interests of higher road speeds), but in the main complicated transmissions are unnecessary on light tanks. They are of far more use on heavy tanks where the forces involved in swinging the vehicle are immense. On VK 30.01 Henschel came up with their L. 320 C, double-differential steering system, which gave the driver the option of three radii of turn.

Although it has no direct bearing on the

BELOW VK 30.01 (H).

Tiger story, mention should also be made of a projected 65-ton tank – VK 65.01 – announced at about the same time as VK 30.01. As usual the hull design was offered to Henschel, while Krupp was placed in charge of turret design. In this case armour protection was to be the predominant feature, up to 80mm at the front, even at the expense of armament. Drawings show a tank looking like a scaled-up Panzer IV, but with torsion bar suspension and overlapping road wheels of much smaller diameter. Regarded as too bulky to travel by rail, Henschel came up with an idea that involved breaking the tank down into three parts to facilitate transportation. On arrival at its destination it was estimated that it would take three weeks to put the tank back together again and this was regarded as a serious handicap. However, the final blow resulted from experience in France in 1940 where road bridge limitations meant that any tank over 30 tonnes could be a tactical liability. Priority dwindled after that and although a trial hull is said to have been completed, by then interest had switched to another new design, VK 45.01.

Meanwhile, there was renewed interest in improved armament and Krupp received instructions to develop a turret for a 10cm gun. Henschel was told that their VK 30.01 chassis should be adapted to take this new gun and turret with the proviso, already mentioned, that an upper weight of 30 tonnes was the limit. This proved impossible, particularly when frontal armour was increased to 80mm, so the new design was designated VK 36.01, but even this proved to be optimistic. On instructions from a higher authority the proposed gun was now to be a 75mm weapon with tapered bore, which would fire a high-velocity tungsten anti-tank round, and armour was to be increased on the hull to 100mm at the front and 60mm on the sides. Even on paper the combat weight was now calculated at 40 tonnes and a desperate shortage of tungsten killed off the gun. In the event, VK 36.01 never developed into a complete tank either, although it played its part in an ongoing story.

The rivals

The appearance of Dr Ferdinand Porsche on the scene in 1939 as chairman of the *Panzerkommission* both complicated and expedited matters. Although Porsche was associated with military vehicle design during the First World War, he does not appear to have been involved with tanks; even so anyone who had studied his work would know that they could expect original solutions to any design problems. In his new role with the *Panzerkommission* Porsche was obviously aware of existing programmes, and seems to have been concerned about the application of mechanical transmissions and steering systems to what were then regarded as heavy tanks. Taking advantage of his position, Porsche

arranged for his firm to obtain a contract for a rival design in the 30-tonne class, VK 30.01 (P), which was also known by the people at Porsche as the Type 100, or Leopard.

Based on previous experience with heavy artillery tractors in the First World War, Porsche advocated a form of petrol-electric drive that obviated the need for any kind of transmission and a form of external suspension using short, longitudinal torsion bars – two features that were characteristic of many of his subsequent designs. The Type 100 was powered by a pair of air-cooled V-10 engines driving a pair of dynamos situated at the rear end of the tank, powering electric motors lined up with the drive

ABOVE VK 36.01 (H) – Albert Speer, Reich Minister for Armaments and Munitions (on the right), is in the driver's position.

LEFT VK 36.01 (H) – Ferdinand Porsche hangs on while Speer drives.

ABOVE In this photograph the Porsche design is fitted with the Krupp turret.

sprockets at the front. Only one running hull has ever been seen, fitted with a simple dummy turret. Contracts for real turrets to mount the 88mm L/56 gun had been let to Krupp, but production was overtaken by events and none were completed. However, as we shall discover, the basic design of the turret would ultimately be seen on the Tiger, so this one-off prototype has a place in the Tiger story.

There is one matter concerning this tank that might be mentioned in passing – the weight. In one source it is given as 30 tonnes, but it is almost unheard of for a newly designed tank to meet its specified weight so exactly, and this has been particularly so in the case of petrol or indeed diesel-electric types. Simple as they might be in terms of power transmission, as Dr Porsche knew, the combination of engines, dynamos and electric motors usually

resulted in a prohibitive weight penalty when compared with conventional systems. Although the tank was never developed any further, Porsche regarded the work undertaken as valuable, particularly in respect of data on the performance of the drive train and experience with air-cooled engines in tanks.

VK 45.01

It would appear that by the spring or early summer of 1941 earlier prohibitions on weight and fears over the loading capacity of bridges had gone by the board. The view, both by Hitler and his technical entourage was that thicker armour and more powerful guns would be essential for future tank warfare and a new project was initiated as VK 45.01 for a tank in the 45-tonne class; both Henschel and Porsche were contracted to produce prototypes under the designations VK 45.01 (H) and VK 45.01 (P) respectively.

The matter was now seen as urgent so it was here that all the work done previously, and outlined above, bore fruit. Even so it is worth bearing in mind that work was still proceeding on the other designs, albeit with reduced urgency as the programme to develop VK 45.01 came to dominate.

As far as Henschel were concerned the prototype VK 36.01 provided much of the chassis design, particularly the suspension, while Maybach came up with a new V-12 petrol engine, the HL210 P45 which could deliver up to 650 brake horse power. The matter of

BELOW VK 45.01 (H)

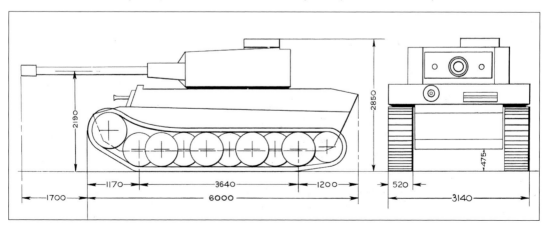

transmission was also evolving. VK 45.01 was provided with a Maybach-Olvar hydraulic pre-selector gearbox offering eight forward and four reverse speeds. This was linked to a new double differential steering system in such a way that the driver had a choice of two radii in any gear, but with the gearbox disengaged the driver could effect a neutral turn, which meant that the tank could spin around in its own length, one track going forwards and the other in reverse.

Thereby hangs a tale. Although it takes us a little bit ahead of the story it is worth noting that in May 1943 a driver's handbook for a Tiger was picked up by British investigators in Tunisia and sent back to Britain for translation, after which it was circulated among interested parties. One of these was an engineer named Ewan McEwen who, having studied the book, recognised the system.

In 1938 the War Office had asked for work to begin on two prototype Heavy Cruiser tanks; heavy in this case indicating tanks in the 20- to 30-ton class. The heavier of the two, A.14E1 was designed in conjunction with the London, Midland and Scottish Railway Company but does not concern us here. A.16E1 does; it was designed in conjunction with Nuffield Mechanisation and Aero Ltd, featured Christie suspension and was powered by a Nuffield Liberty V-12 engine rated at 414hp. At about this time the Department of Tank Design at Woolwich, in association with Dr Henry Merritt, was developing a sophisticated double-differential tank steering system incorporating a seven-speed gearbox designed, in conjunction with the DTD by Maybach Gears Ltd, the British branch of the famous German company. One of the directors of the British company was a German national named Ernst Schneider, who returned to his homeland shortly before the war. It is generally regarded as no coincidence that the steering arrangements for the Tiger appear to have evolved from this British prototype. It has to be said, as McEwen appreciated, that the original Merritt–Maybach design was big and over-complicated and he assumed that the German designers simplified it, as indeed they did, and it is equally true that Henry Merritt's later design, the Merritt–Brown as used in many British tanks, also eliminated a lot of the complications embodied in the prototype.

Of course the bridging problem had not gone away, just because it had been ignored. Indeed now that Hitler's armies were on the verge of moving East the problem had escalated. If anything rivers were wider and bridges even flimsier in Russia than they were in the West, and tanks were getting heavier. As a result the new tank would have the ability to deep wade through waterways, up to a depth of about 4.5m, built into the design. This meant that a great deal of extra work had to go into the design and construction of the new vehicle. Water had to be kept out of the engine and crew compartments but admitted to the isolated radiator compartments albeit with the cooling fans turned off. A form of breathing

ABOVE VK 45.01 (P) – the Porsche design on trials, fitted with a dummy turret.

BELOW The trial at Haustenbeck near Sennelager. Here the Tiger is fitted with a *snorkel* in the commander's cupola.

apparatus or *schnorkel* device was fitted, in the form of a long telescopic tube which could be extended before the tank went diving.

Another feature considered at this early stage was a sort of hinged shield, fitted to the front of the hull; normally laid on top of the front glacis plate, it could be dropped down on hydraulic arms primarily to shield the tracks from frontal attack on level ground. It is probably also worth pointing out at this stage that the designation VK 45.01 was almost meaningless – the Tiger would tip the scales at around 57 tonnes.

Yet that was not the entire story by any means. Based on their earlier experience with the Type 100 the Porsche people also embarked upon an entry for the Tiger programme with their own design, VK 45.01 (P) or Type 101. The shape of the hull was derived from the Type 100 but with thicker armour at the front and wider tracks to reduce the ground pressure because it also weighed 57 tonnes. In essence Porsche stuck to his guns, with up-rated V-10 air-cooled engines, petrol-electric drive and his favourite longitudinal torsion bars. However, there were changes, the electric motors were shifted to

the back and lined up with rear sprockets, a rare feature on German tanks, a new style of steel-rimmed road wheels and for some reason toothed idler wheels at the front. However, the most important feature by far was the turret. Designed by Porsche but constructed by Krupp and clearly derived from a version originally planned for the Type 100 it featured a flat roof and vertical sides encompassed by a huge, horseshoe-shaped outer shell. Across the front a massive, cast mantlet contained the powerful, almost legendary 8.8cm *KwK* 36 gun (*Kampfwagen Kanone* (*KwK*) or fighting vehicle cannon). This was the final piece of the puzzle, with this turret mounted on the new Henschel hull one had, in effect, the classic Tiger.

Just to round off the story of the Porsche contestant, just five examples were completed but they experienced endless trouble with the drive train and further production was stopped. About a year later, in the early summer of 1943 the basic hull design was revived, including the same suspension but now fitted with a pair of Maybach HL 120TRM water-cooled engines, moved to the front. It featured a new, rigid superstructure at the back and mounted the big PaK 43 8.8cm gun (*Panzerabwehr-Kanone* (*PaK*) or anti-tank gun) with the long L/71 barrel. In this form as the *Ferdinand*, later *Elefant*, it proved to be a very effective tank destroyer.

Production

A prototype of the new Tiger tank was demonstrated to Hitler on his birthday, 20 April 1942, near Rastenburg in East Prussia. Contracts for production, in batches, had already been placed with Henschel und Sohn for the assembly work with hull armour supplied by Krupp, in Essen, subcontracted to Dortmund Hoerder Huttneverein (DHHV) for a while. Wegmann Waggonfabrik which, like Henschel, was primarily in the railway business and also based in Kassel, was responsible for supplying turrets while many other firms were involved in supplying components; Maybach of course for the engine, Adlerwerke in Frankfurt for the transmission, Henschel themselves for the steering gear and two firms, DHHV and Wolf Buchau, supplied the 8.8cm guns.

It should not need saying that, initially at least,

BELOW Ferdinand Porsche (left) and (in uniform) Albert Speer, pictured on the rear engine deck of the Henschel hull, which is fitted with a ballast turret.

there were endless mechanical problems to be overcome that slowed down production. This was endemic in most new machines fresh off the production line, but particularly so in the case of a heavy tank such as the Tiger, which was pushing the limits of technology and material strengths in just about every direction. As a consequence production was delayed and, even when the first Tigers were delivered to the Army on the Russian front near Leningrad in August 1942, three out of the four were found to be inoperable.

The protective front shield, seen on the first prototype at Henschel in April 1942, was dropped from the production vehicles. This resulted in a redesign of the track guards and with operations in a dusty environment in mind, extra air filters by Feifel were fitted to the rear of the hull at each corner. Indeed there were many more modifications introduced from time to time, although many relate to the later period of Tiger production. Records show that in total 1,354 Tiger tanks were produced between July 1942 and August 1944, by which time a newer and larger model, the *Ausführung* B (better known as the King Tiger) took over.

As mass production programmes go this was hardly impressive, particularly when compared with American or Soviet production and this can be blamed to some extent upon the effects of Allied bombing and shortages of raw materials. However, it is also not unreasonable to point out that, typically, the German engineering ethos of experimenting

ABOVE The Tiger prototype featured a hinged shield, which could be lowered in front of the hull to protect the tracks. It was dropped from the production model.

Fahrgestell Nummer and the *Waffen Prüfungsamt*

Each armoured fighting vehicle made was given a chassis number or *Fahrgestell Nummer* – abbreviated to *Fgst Nr.* This unique number identified where the tank came in the production series. Henschel was issued seven separate contracts for the production of the Tiger starting with the *Fgst Nr* 250001.

During the manufacture of the tanks the *Waffen Prüfungsamt* 6 (Weapon Testing Office 6, the Army Weapons Office department responsible for tank design, abbreviated to *Wa Pruf* 6), issued amendments to the designs. These followed reports from units using the vehicles in the field. The rushed programme and failure to allow adequate testing led to many initial faults and consequent amendments in the vehicle's design. Changes were also made to the vehicle to speed up production processes or in response to material shortages, an example being the introduction of new road wheels that saved on the use of rubber.

There was an inevitable delay from the time of issuing amendments to when the design change appeared on vehicles leaving the factories. Old parts were used up first, or in some cases were left at the back of the store when new parts arrived and took the prominent front position. The old parts were then used only when the new were exhausted or supplies interrupted.

Many of the changes to the tank were small and subtle; others, like the new design of road wheels were much more visually obvious. From July 1943 with *Fgst Nr* 250391, the turret of the Tiger went through major changes including a new commander's cupola.

Manufacturers were issued with a three-letter code system so the origin of components was known but captured equipment would not reveal the maker and therefore the location of manufacture, which could become a target for bombing. This lettering system can help trace the originators of parts as many of these codes have now been identified (bwn for example is Krupp). Acceptance marking by the *Waffen Prüfungsamt* can also be found on a lot of parts. The accepted item is stamped with a small eagle above WaA and/or a three-digit number that identified the individual inspector.

David Willey

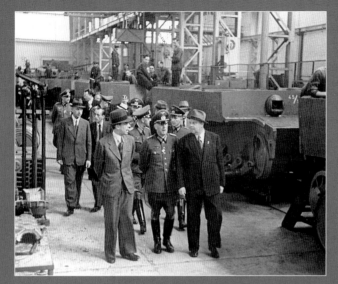

LEFT A factory visit to the Henschel works at Kassel.

Building Tigers – David Willey

The hulls of the Tiger were assembled at Henschel's factory in Kassel. Henschels had to move railway locomotive production from one of its massive production shops (Shop 3) and shared another (Shop 5 where locomotives were made on the left side, tanks on the right). On average, there were always 18 tanks in the hull assembly workshop and 10 on the final assembly line. The final assembly had two lines of vehicles and one side cranes would manoeuvre hulls down the nine *Takts* or stations. Henschel had to employ many new workers and a brand new accommodation centre was built by the Organisation Todt near the plant test track. In total Henschel employed 8,000 workers who carried out two daily 12-hour shifts (the night shift always less productive than the day).

Wegmann Waggonfabrik AG in Kassel assembled the turrets and delivered them to Henschel's for fitting. Wegmann employed 1,200 workers in two ten-hour daily shifts. Many workers were foreign labourers but at no point were prisoners-of-war allowed to work on war material. In turn many suppliers of sub-assemblies and components were spread around Germany, suffering from 1943 onwards the effects of increasing Allied bombing of their factories and rail infrastructure.

It must be remembered that the production of vehicles in Germany prior to the Second World War was far from the mass production Ford assembly line we usually imagine. Vehicle production, if not

LEFT The production shop inside the Henschel works.

quite a cottage industry, was carried out by many small workshop concerns and statistically Germany was behind France, Britain and America in vehicle ownership. In the planning and preparation for war, General Adolf von Schell organised the curtailing of the production of numerous small runs of vehicles by differing manufacturers into a standardisation of models and categories of vehicles. This became known as the Schell programme. However, the manufacture of tanks was still a labour-intensive activity and a huge and complex undertaking.

The production of Tigers had a number of initial problems but by December 1942 the production process was in full swing with monthly totals of vehicles increasing from 30 to peak production in April 1944 of 104 vehicles. Bombing in October 1943 slowed production at the Henschel plant. This led to the conversion of 18 *Panzerbefehlswagens* (or command vehicles) back into standard gun tanks to make up the required numbers. Henschel ran down production of Tiger I in the summer of 1944 as the production of SdKfz 182 or Tiger II increased. (SdKfz was the abbreviation for *Sonderkraftfahrzeug* or special-purpose vehicle.)

Krupp made 537 and DHHV 758 armoured hulls and turrets for Wegman and Henschel to assemble. The remaining hulls and turrets to make the total production of 1,354 were made up from refurbished battle-damaged vehicles with some new components.

Engines were delivered from Maybach, production being interrupted by a bombing raid in April 1944 that stopped manufacture until October 1944. Auto Union also made the HL230 P45 engine and 219 engines for the Tiger I, which were delivered in 1944.

Production of certain components seemed barely to keep up with the manufacturing programme and gave little leeway for spare-part production. In December 1944 it was reported that there was only one spare steering unit and one spare gearbox for every ten tanks in service.

and developing to near perfection was also to blame. Many authors have already brought out the diametric difference between engineers such as Dr Erwin Aders, the methodical chief engineer at Henschel and the mercurial Dr Ferdinand Porsche and it is worth remarking that while Aders complained bitterly about the rushed procedures and short cuts that spoiled the perfection of his new design, none of the projects emanating from the Porsche organisation, except maybe the Volkswagen car, ever competed successfully against rivals and, with the exception of the much modified *Elefant*, ever entered service.

From factory to front line

Final assembly, including painting and preparation for service, was done at the Henschel plant in Kassel. Tanks were sprayed in a red oxide primer and then a base coat of the ruling colour of the day, either Panzer Grey or the golden brown *Dunkelgelb,* which subsequently became standard. Final

LEFT A factory-fresh Tiger is issued on transport tracks. This tank is fitted with *Feifel* air filters, anti-personnel mine dischargers on the hull corners, and brackets for holding spare track links on the turret side.

ABOVE **A Tiger from the 501st Heavy Tank Battalion boards a transport for Tunisia.**

Trapani in Sicily at the end of the month. From here, early in March tanks from the 1st Company were shipped across to Tunis by Siebel Ferry, three at a time. Here they were reinforced by 11 Tigers of what was left of the 501st Heavy Tank Battalion, which had been in Tunisia since the previous November. Their arrival was duly reported in a British newspaper, the *Daily Mail* of 30 January 1943 under the headline '62-ton German Tanks Arrive' and then went on to boost them as 'monster' tanks, 'land battleships' even, with 'seven-inch armour plating'. How this was supposed to improve the morale of the British soldiers who read it is hard to imagine.

The organisation of a Heavy Tank Battalion at this time, at least in ideal circumstances, was a command platoon and three fighting platoons. The command platoon consisted of one Tiger and two Panzer IIIs; each fighting platoon had two Tigers and two Panzer IIIs. Our Tiger, serial number 250122, was issued to the commander of No 3 platoon so the turret was adorned with the number '131', in red on each side. This signified the Company, the Platoon, and the place of that particular tank within the Platoon.

On 17 March the reinforced company set off on a 400km road march from Tunis to the area Sfax–Maknassy. Twelve Tigers arrived out of a possible total of fourteen, which is not bad under the circumstances, although it does help to emphasise the relative unreliability of these

camouflage arrangements were left to units in the field.

Since new tanks left the factory by rail they were fitted with narrow tracks with various other fittings temporarily removed. On arrival at the end of the line these parts were restored, extra wheels and wider tracks fitted which spread the weight for driving over soft ground into action.

Enter Tiger 131

One tank in particular, serial number 250122, rolled out of the Henschel works in February 1943. It was issued to the 504th Heavy Tank Battalion and moved by rail to

RIGHT **This Tiger, being loaded onto a railway flat, has the narrower transport tracks fitted.**

new tanks – a mixture of untested and over-stressed components and inexperienced crews.

We next hear of 250122 (Tiger 131) on 19 April 1943 when two Tigers and other tanks under *Oberleutnant* Witt took part in Operation *Fliederblute* (Lilac blossom) in an area known as the Djebel Djaffa. They were not alone. The 48 RTR, which had also arrived in Tunisia in March 1943, but this time direct from Britain, had just been moved by rail to the same area and went into action. Equipped with Churchill tanks the British regiment moved against German positions in the hills at Medjez-el-Bab and the two forces met, in a muddled engagement on the 21st. As a result the British lost two Churchill tanks but a couple of lucky shots had also disabled our Tiger. Quite what happened, quite how much damage was done to the German tank is not easy to assess. Evidence of the damage remains but that does not tell us the effect this had on the tank itself. The German War Diary gives its own take on the event: '*The crew members of TIGER "131" panic and abandon the tank after two harmless hits from a CHURCHILL.*'

The original crew have never been traced so the truth of this may never be known, but the tank itself, still in potential working order aside from a jammed turret, was now in British hands. A contemporary report claims that it was recovered and repaired by 104 Army Tank Workshops, REME, using a Caterpillar D6 tractor as the recovery vehicle. Once it was back in working order the tank was used to make a recognition film in a location that the British called Redcap Alley, and then it was moved into Tunis itself on 24 May 1943 where it was subsequently shown to His Majesty King George VI, Prime Minister Winston Churchill and Foreign Secretary Anthony Eden.

A Tiger in full working order was considered quite a prize and it was planned to ship it back to Britain, although this was no simple matter. The tank was driven under its own power to the harbour at La Goulette and was ferried aboard a Tank Landing Craft to Bizerta on 3 August 1943. From Bizerta it was shipped aboard the SS *Empire Candida* to Bone on 9 August and from there on board the SS *Ocean Strength* (Captain William Rickard) for Glasgow where it docked on 8 October 1943.

The Tiger was destined to be delivered to the Department of Tank Design in Surrey, where it arrived on 20 October 1943, but the following month saw it displayed on Horse Guards Parade in London – coincidentally the same location where two captured German A7V tanks were placed on public display in 1919.

The Tiger's precise moves after that are not well recorded. It was subjected to mobility trials on the ranges in Surrey, to gunnery trials at Lulworth in Dorset and appears, at least for a while, to have joined a touring road show nicknamed 'Martel's Circus', inspired by Lieutenant-General Sir Giffard le Quesne Martel, which toured military camps and population centres throughout Britain with a selection of Allied and captured enemy tanks for familiarisation.

As a result of the action at Medjez-el-Bab, where the Tiger was captured, a troop leader in 48 RTR, Lieutenant Peter Gudgin, was wounded when his Churchill tank was knocked out and he was subsequently invalided home to Britain. Once he had recovered he was posted to the School of Tank Technology, a branch of the Military College of Science at Chertsey, and by a bizarre coincidence found himself responsible for writing intelligence reports on the very tank that caused him to be in Britain in the first place.

The Tiger itself, once its travels were over, returned to Chertsey for a thorough examination. This involved not only taking the tank to bits to measure every component, but also compiling

BELOW The crew of Tiger 131 in Tunisia. Note the welded front step and hand grip, and that only two smoke dischargers are present on the left turret bank.

ABOVE Tiger 131 on 22 April 1943 in the position it was found. The helmeted figure is looking back in the direction of the attack by 48 RTR.

ABOVE The Tiger in captivity. Careful study of the original photographs that were taken soon after capture reveals a two-tone paint scheme on the tank.

RIGHT The rear of the tank shows blast damage from high-explosive rounds. An example of the *Panzergranate (Pzgr)* 39 round – the standard German armour-piercing round – has been placed beside the track.

FAR RIGHT This image shows further damage on the exhaust covers at the rear of the tank. The '122' painted on the rear left *Feifel* air filter relates to the '*Fgstr.*' or unique serial number 250122. Closer inspection reveals a two-tone camouflage scheme on the turret.

RIGHT The right-hand side of the vehicle. Note the padlock on the rear turret bin, which is still in place today. Why someone chalked 'G' or '6 LOOK' on the side is unknown.

reports on all the relevant parts. For example Armstrong-Siddeley Motors undertook a very thorough metallurgical evaluation of the gearbox and steering system, while the staff at Chertsey removed the engine, chopped it open to see how it was designed and in the process rendered it useless.

In fact this process of evaluating the big tank and writing it all up overran. Following the invasion of Normandy in June 1944 many more modern German vehicles arrived at Chertsey and the Tiger gradually slipped down the priority list. As a result the full report was never finished and the tank remained in a semi-complete state. It was officially handed over to the Tank Museum on 25 September 1951; but that is another story.

ABOVE King George VI is shown Tiger 131 in Tunis by Brigadier Cook.

LEFT The Tiger is displayed on Horse Guards Parade in central London. It had been suggested that the tank should be a 'gift' to the Prime Minister from the 1st Army, but at the time of the presentation of the tank in London, Churchill was in Marrakesh recovering from a heart attack.

Tiger variants – David Willey

O nce the production of the VK 45.01 (H) model had been agreed, there were considerable changes to the Tiger during the production period, but no official model variations or *Ausführung* letters were issued to identify significant changes. This has led post-war commentators to give a general description to vehicles as being early, mid- or late production models.

Sturmmörser Tiger

Eighteen Tigers were converted into a rocket-firing mortar carrier. The vehicle was conceived in August 1943 at a time when siege warfare was still possible on the Eastern Front and the memories of Stalingrad fresh. ALKETT presented a prototype vehicle in October 1943 but it wasn't until August 1944 that work began converting gun tanks into the mortar carriers and work was completed by December. Essentially the turret and top decking of the Tiger I was removed along with the front driver's plate and a large boxed structure housing the 38cm RW61 rocket-assisted mortar was placed on top.

Each vehicle carried twelve 38cm rocket missiles that weighed 325kg each. A crane was used to manoeuvre the rockets onto the

ABOVE The *Sturmmörser* Tiger. This captured example is fitted with a 38cm RW61 rocket-assisted mortar and shows the jib for manoeuvring the massive rounds that it fired. Also shown is the *Zimmerit* coating on the hull, evidence of the vehicle's conversion from a *Zimmerit*-coated gun tank.

RIGHT The projectile fired by the *Sturmmörser* Tiger. Twelve such rounds could be carried by the vehicle.

BELOW Germany, 1945 – a British-crewed Sherman recovery vehicle passes a captured *Sturmmörser* with American troops perched on the top.

BELOW The mortar tube from the *Sturmmörser*. Marked as 'R19', 'Piece 19', the rifling inside the barrel to impart spin to the round can be seen.

vehicle and a winch located on the roof was used to transfer rockets from storage racks onto a roller assembly to load into the firing tube. The missile, when fired, was assisted in flight by rocket propulsion and some of the propellant gasses were vented forward through tubes around the barrel. The gun could propel the missile 4,600m and the size of detonation was very impressive but the nature of the fighting had changed by the time the *Sturmmörser* came into service. They were used in the Warsaw uprising and the Ardennes campaign and in the defence of the Ruhr but with so few vehicles made, they had little impact on the outcome of any battle in 1945.

Three *Sturmmörser* companies were created, 1001, 1002 and 1003, to man the vehicles, each taking a crew of six as two loaders were required. One captured example was returned to Chobham for evaluation. After the war it was broken up and its gun has made its way to the Tank Museum after many years at the teaching academy at Shrivenham.

Field modifications

At least one Tiger was found abandoned in Italy that started a rumour that a recovery variant of the tank had been improvised. In fact the tank was being used by *Schwere Panzer Abteilung* 508 (508th Heavy Tank Battalion) as a test bed for an improvised mine-clearing device. The gun had been removed and a boom and winch mounted atop the turret. This was intended to carry charges in front of the tank to be lowered and detonated over potential mines.

TOP The geared winding mechanism on the back of the Tiger's turret. The device was almost certainly manually operated.

LEFT The converted Tiger that led to a flurry of reports suggesting it might be a recovery version. In fact it was a conversion to detonate mines.

ABOVE Part of the winch arrangement placed on the turret of the Tiger that was converted for mine-clearance operations in Italy. The cupola is of the second design.

'Churchill arriving here tomorrow. Will want
to see Tiger and probably drive in it.'

Order issued to Major Douglas Lidderdale, REME
22 April 1943

Chapter Two

The Travels of Tiger 131

David Willey

The action around Djebel Djaffa on 21 April 1943 was just a small action on one day of a very long war. The significance of the capture of Tiger 131 was only recognised in certain quarters. Before looking at the restoration of Tiger 131 it is interesting to examine in more detail the efforts that went into recording the tank, getting it to Britain and what happened to it there.

OPPOSITE Tiger 131 on 22 April 1943. Materials taken from the interior are scattered over the front of the glacis plate. From left to right: bolt cutters, *Gurtsack*, optics boxes, disposable oil, smoke pot, and *Gurtsack* lid. In front of the vehicle can be seen canvas from the stowage bin and what may be tools. Closer scrutiny reveals that below the mantlet is evidence of battle damage, along with the shell-struck lifting eye and smashed loader's hatch.

high-explosive round bursts behind his position and as he pans the camera back – over a sheltering helmeted soldier – smoke and earth can be seen falling and the closeness of one of the knocked-out Churchill tanks from 48 RTR is clearly visible. This tank was probably Lieutenant Peter Gudgin's, who was later to be involved in the analysis and report on the Tiger back in Britain. Still shots of the tank were also taken at this time – the low angle with grasses and thistles in the foreground, not included as one might suppose for artistic effect but taken by the cameraman, sensibly keeping his head down against shell fire.

Whether the Germans were specifically trying to destroy the vehicles left after the action or just generally shell the British positions is not known.

Later the same day a closer inspection of the battlefield allowed a summary to be written for

For 48 RTR, the importance of the action on 21 April 1943 was that it was their first time in action, and the attack was considered a disjointed affair between infantry and tanks. Lessons would need to be learned and the action was cited in a War Office Training publication, *Notes from Theatres of War No 16*, as 'an example of poor cooperation'.

On the day after the action, the Tiger was photographed and filmed. A short piece of film footage taken that day shows the vehicle from one of the trenches immediately in front of the tank. As the cameraman films, a

the 48 RTR War Diary and further photography to be taken of the tank. This included a shot where soldiers from the East Surreys can be seen inspecting the tank along with the Commanding Officer of 48 RTR, Lieutenant-Colonel G.H. Brooks. A Churchill from 48 RTR can be seen in the background as well as a Dingo scout car from Regimental Headquarters. Images show items removed from the vehicle on the glacis plate, including a pair of German bolt cutters, two stowage boxes, a silvered card disposable oil can, and a smoke pot from one of the dischargers.

Further film footage taken at a later date but at the same spot (marked as Point 174 on British military maps) shows traffic moving past the Tiger and heading down the reverse side of the slope from the tank. This indicates that the vehicle's position was (as you faced the tank) just to the right off a track on the brow of the defended position.

The recovery of the tank was entrusted to 104th Army Tank Workshops and Major Douglas Lidderdale of the Royal Electrical Mechanical Engineers was ordered to supervise the operation. On 7 May, over two weeks after the action, he arrived on the scene with a D6 tractor. Reg Whatling, one of Lidderdale's crew ensured the vehicle had no booby traps before a fuller inspection was made. Lidderdale later wrote 'I can confirm on the 7/5/43 the turret was jammed', he added 'by a burr on the armour protecting the turret ring'. In a later account he stated: 'The Tiger's radio was shattered by the round which split the weld on the top plate, it is presumed to have incapacitated the driver and front gunner.'

The early photographs also clearly show the cracked loader's hatch, which was subsequently replaced. To be damaged in such a manner the hatch would have had to be open. Could this impact have also caused injury to a crew member?

The vehicle was checked for oil, fuel and water, all were present but the water level was a little low. The vehicle was pulled by the D6 tractor to Scorpion Corner, Redcap Alley nearby to Medjez-el-Bab and, en route, was momentarily filmed again by an official photographer.

At Redcap Alley the tank was run for the

cameras and filmed from a number of angles to make a recognition film called *The Tiger* (Film B406), produced by the Directorate of Army Kinematography. The film starts with a ticker-tape opening credit 'Flash – Captured on Tunisian Battlefield' and goes on to show the Tiger being driven through a haystack and advancing over a crest, as well as demonstrations of the deep-wading *schnorkel*. The narrator gives advice on the key recognition features of the tank such as the boxy hull, offset cupola and driver's vision port on the left. Comment is also made on the 'tail-up' look of

the tank as if the turret is too heavy for the 'flat deck' of the hull. The film ends with what was hopefully comforting images of blown-up Tigers for the Allied soldiers viewing the film.

Lidderdale recalled the tank ran well for the filming. However, when the tank was ordered to Tunis, it was run for a much lengthier period leading to overheating and steam appearing from the engine bay. Stopping the vehicle to inspect the cause, Lidderdale's previous experience as an engineer for Leyland came in handy. Water on the floor of the engine compartment led him to identify a 'failed rubber sealing ring between the wet cylinder liner and aluminium cylinder block/crankcase unit'. Later, Lidderdale speculated whether this overheating may have caused the German crew to abandon the tank, too, during the action on 21 April. For

a short period the tank appeared to run well, but during an extended period of running it overheated. With fire incoming and an overheated engine, perhaps the crew decided to abandon a then temporarily immobile vehicle?

Keeping this first, secured Tiger intact and in potential running order was recognised as of paramount importance. Despite requests to see the vehicle running for dignitaries such as Alexander, Churchill and the King, Lidderdale saw to it that the Tank ran no more in Tunisia so the engine was not put at risk of overheating.

In Lidderdale's papers a copy of an order remains:

'To Major Lidderdale from Lt Col Davis. Churchill arriving here tomorrow. Will want to see Tiger and probably drive in it. Report to

Churchill handles an 88mm round during his inspection of the tank in Tunis.

racecourse earliest. If required we may get drivers to bring it here but this not decided yet.'

Lidderdale was present when the tank was shown by Brigadier Cook from 1st Army Headquarters to King George VI. He recalled the tenseness of Cook, as during the visit Cook had a fraught conversation with James Grigg, Minister of War, about the effortless ability of German tank guns to 'pee through' the armour on British tanks.

Instructed to gather spares to help assist in the running of the tank and to test materials, such as armour plate, Lidderdale acquired some items from the blown-up Tigers at Hunt's Gap but the Tiger repair workshop at Manouba, just outside Tunis, had already been taken over by the US Army who were also sourcing spares for their Tiger, donated to them by British forces. The following items were listed as ready to ship to the UK by Lidderdale: '1 Tiger Tank complete, 11 boxes containing spares, 4 pieces of armour plate, 2 lengths spare tracks and 16 bogie wheels'.

Orders were received to return the tank to the UK and the vehicle and spares were ready to depart in late June, but it was not until the end of September that the tank was loaded at the Port of Bone. Meanwhile, the 104th

ABOVE Brigadier Cook shows King George VI the tank. The shield of 1st Army and the top half of the 21st Army Tank Brigade diabolo can be seen on the front plate.

LEFT Destroyed Tigers at Hunt's Gap.

RIGHT The driver's position inside the tank. Damage from the shell strike can be seen in the top centre of the image.

BELOW The smashed radios have been removed from the shelves in the centre between the driver's and bow machine gunner/ radio operator's positions.

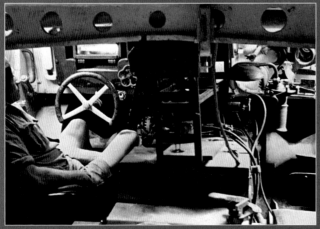

Battle Damage – David Schofield

David Schofield lectures in forensics and crime scene science. He was invited to survey the damage on Tiger 131 and here are some of his findings:

Tiger 131 still shows effects from its time in combat. Some damaged items were replaced soon after capture, the loader's hatch in the turret, for example, the original of which shows clear damage on capture. Photographic evidence shows the interior damage above the driver's and radio operator's positions as being more extensive than we might now assume and Major Lidderdale who recovered the vehicle clearly states the radios were smashed. The depression here was repaired at Chobham and a D-shaped metal insert welded in place.

In summary the vehicle was hit by at least three 6pdr [57mm] anti-tank rounds, at least 30 0.303in [7.7 mm] rounds and approximately 13 shrapnel hits of various sizes.

The damage

[All references to position are from an observer's perspective looking at the front of the tank]. A forensic examination of this vehicle, based on the above, shows:

ABOVE Scarring from shell damage around the mantlet.

RIGHT The shell impact point beneath the gun mantlet. The D-shaped insert in the hull decking in the foreground is a later repair after capture.

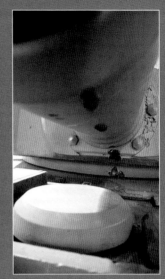

ABOVE Following the trajectory of the round beneath the barrel to the mantlet.

The vehicle had at least eight separate identifiable attacks on it. Of these at least five certainly happened during its last combat. It is likely that all occurred at this time.

The attacks may have caused injury to up to four of the five-man crew, the commander, loader, hull machine gunner and driver.

Attacks

The attacks consisted of:

1 A 57mm solid shot round hitting the underside of the main gun 4 degrees off from the gun to the right and continuing to strike the barrel in three places, then the gun base and then the base of the turret and the hull top plate.
2 A 0.303in-calibre machine-gun burst raking the gun and mantlet in the same area as strike 1 above. These follow a slightly higher trajectory than the anti-tank round. Some rounds hitting the top of the gun and mantlet.
3 A 0.303in burst striking the *open* commander's hatch, vision slit and cupola.
4 A 57mm round hitting the right side turret lifting stud and deflecting upwards and slightly away.
5 A 57mm round through the loader's raised hatch and deflecting down and to the right to hit the edge of the hatch opening.
6 A burst of 0.303in machine-gun fire, from approximately 10 degrees off to the right,

hitting the right hull of the vehicle near the rear.
7 A burst of 0.303in machine-gun fire, from 30 degrees off to the left and slightly lower, hitting the left hull and road wheels.
8 A blast attack from the rear approximately 30 degrees off to the right, hitting cyclones, exhaust and pannier. The latter hit is discussed in more depth later.
9 A potential hit to the driver's hatch [based on photographic analysis] discussed later.

Attacks 1–6 are generally from the front and likely to have been from the Churchill tanks of 'A' Squadron. Attack 7 is from the left and probably from 'B' Squadron. Attack 8 is high explosive and could have come from an HE round from the British or a German artillery attack subsequent to the position being occupied by the British.

LEFT The shell strike on the lifting eye.

Army Tank Workshop had been disbanded but Lidderdale was ordered to keep a select group together to shepherd the tank back. In the waiting period he composed a preliminary report on the tank, copies of which still exist in the Tank Museum's Archives. The tank sailed on the SS *Ocean Strength* with 10,000 tons of iron ore to Glasgow. Here it was collected by Pickfords with a 100-ton Scammell transporter and taken to Chobham, arriving on 20 October. The evaluation of the tank by the Department of Tank Design was to take place at Chobham and gunnery trials were carried out at Lulworth in Dorset. In one of those peculiar synchronicities that life can throw up, the now Major, Peter Gudgin, was tasked with writing the reports

on the armament, the fighting compartment, stowage and the power traverse systems of the tank that had only months before, knocked out his own tank in Tunisia. Major J.P. Barnes wrote the sections on the engine and transmission.

The reports they wrote were thorough, informative and beautifully illustrated and have already merited reproduction in book form.

After delivery of the Tiger, Lidderdale was also posted to Chobham before being posted as the Assistant Director of Tank Design, Special Devices Branch and Armoured Assault Equipment. He saw through the development of a number of the specialised engineer vehicles such as Flails and AVREs for the D-Day landings.

ABOVE Tiger 131 photographed outside the hangar at Chobham with a broken track.

ABOVE RIGHT The Tiger was test-run on a number of occasions – here in muddy conditions. During gunnery trials at Lulworth, a major engine failure occurred that led to a rebuild of another engine with spares brought back from North Africa.

RIGHT A cornucopia of armour at Chobham. The Tiger is marked 'DTD [Department of Tank Design] 3016'. An amazing collection of Allied and captured armour fills the shed, but the Panther and the Ferdinand in the right foreground are both mock-ups (look closely at the tracks of both vehicles).

LEFT Major Lidderdale and his crew in front of the Tiger on Horse Guards. They were among the first British soldiers to wear the newly issued Africa Star medal ribbon. The white marks on the Tiger's front plate are from a poldi hardness test to determine the strength of its armour.

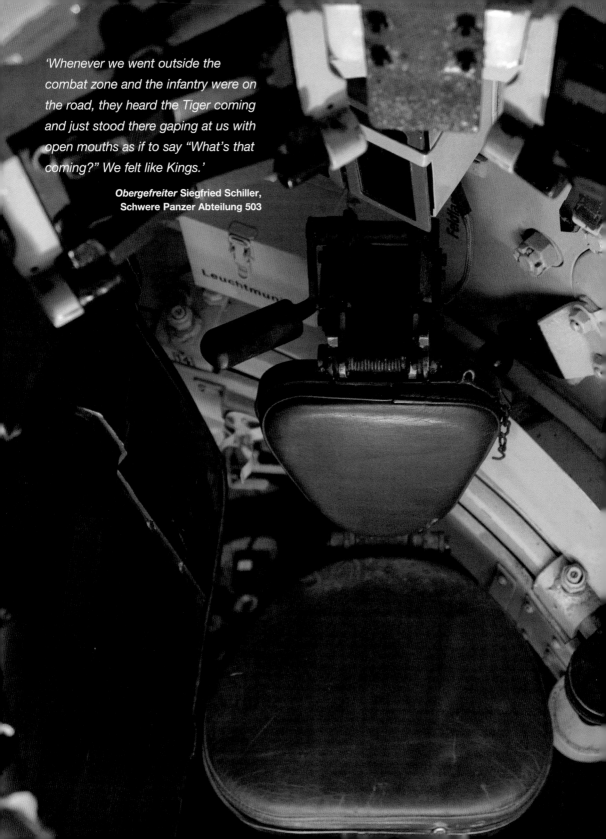

'Whenever we went outside the combat zone and the infantry were on the road, they heard the Tiger coming and just stood there gaping at us with open mouths as if to say "What's that coming?" We felt like Kings.'

Obergefreiter Siegfried Schiller, Schwere Panzer Abteilung 503

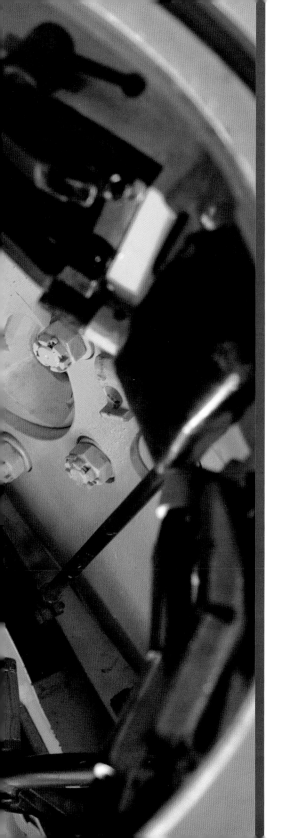

Chapter Three

Anatomy of the Tiger

David Willey

The Tiger was in most areas a tank designed with a traditional layout. It differed from other contemporary tanks by its increased scale in areas such as gun size and armour protection, and by the provision of a deep-wading capability. This chapter allows us to look at some of the key physical features of the vehicles.

OPPOSITE Looking down into the commander's position. The seat could be folded up or it could be stood upon to vary the viewing height from the cupola.

RIGHT **Cutaway drawing of the Tiger I from the School of Tank Technology report.**

HOLDER FOR BOX CONTAINING M.G. GROUND MOUNTING

FOR WATER

SMOKE GENERATOR DISCHARGERS

BALANCE SPRING CYLINDER

M.G. ACCESSORIES

8·8 AMM BIN

MOUNTING FOR WIRELESS SET

8 GUI L

DISC BRAKE DRUM

SHOCK ABSORBER

STEERING UNIT

STEERING WHEEL

DIRECTION CONTROL LEVER

EAR SELECTOR LEVER

DRIVER'S SEAT

STARTER CARB. CONTROL

HAND BRAKE

EMERGENCY STEERING LEVERS

ACCELERATOR

FOOT BRAKE

CLUTCH

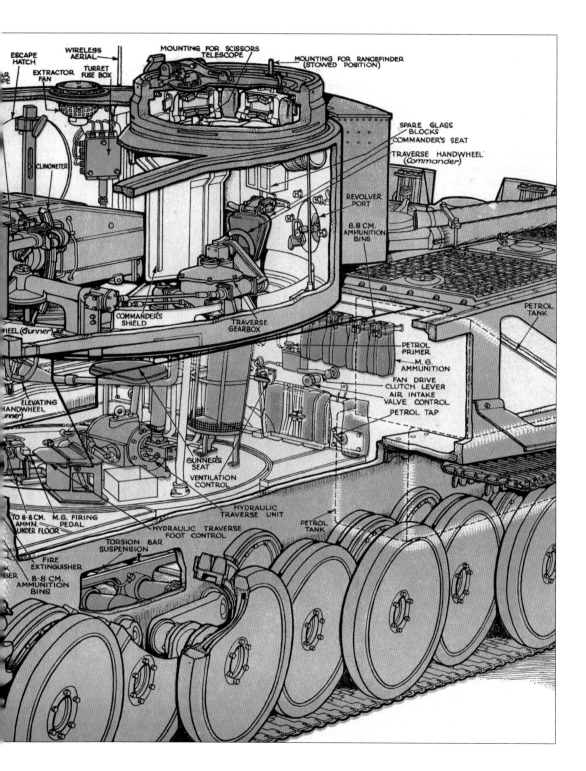

ESCAPE
HATCH

WIRELESS
AERIAL

MOUNTING FOR SCISSORS
TELESCOPE

MOUNTING FOR RANGEFINDER
(STOWED POSITION)

EXTRACTOR
FAN

TURRET
FUSE BOX

CLINOMETER

SPARE GLASS
BLOCKS
COMMANDER'S SEAT

TRAVERSE HANDWHEEL
(Commander)

REVOLVER
PORT

8.8 CM.
AMMUNITION
BINS

PETROL
TANK

WHEEL (Gunner)

COMMANDER'S
SHIELD

TRAVERSE
GEARBOX

PETROL
PRIMER

M. G.
AMMUNITION

FAN DRIVE
CLUTCH LEVER
AIR INTAKE
VALVE CONTROL
PETROL TAP

ELEVATING
HANDWHEEL
(Gunner)

GUNNER'S
SEAT

VENTILATION
CONTROL

HYDRAULIC
TRAVERSE UNIT

PETROL
TANK

TO 8.8 CM.
AMMN.
UNDER FLOOR

M.G. FIRING
PEDAL

HYDRAULIC TRAVERSE
FOOT CONTROL

TORSION BAR
SUSPENSION

FIRE
EXTINGUISHER

8.8 CM.
AMMUNITION
BINS

Chassis from rear.

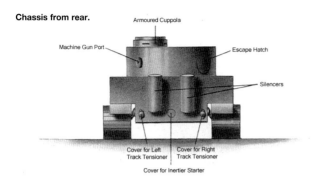

Armoured Cuppola

Machine Gun Port

Escape Hatch

Silencers

Cover for Left Track Tensioner | Cover for Right Track Tensioner

Cover for Inertier Starter

Chassis from below.

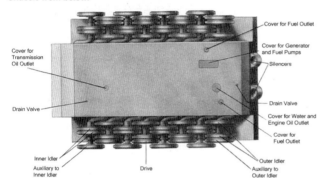

Cover for Fuel Outlet

Cover for Generator and Fuel Pumps

Silencers

Cover for Transmission Oil Outlet

Drain Valve

Drain Valve

Cover for Water and Engine Oil Outlet

Cover for Fuel Outlet

Inner Idler

Auxiliary to Inner Idler

Drive

Outer Idler

Auxiliary to Outer Idler

General layout

The hull of the Tiger was divided into two main compartments; the engine compartment at the rear and the fighting compartment at the front. Between the two was a firewall. The rear engine compartment housed the engine in a tight central bay with two fuel tanks, the radiators and fans housed in side panniers. From the inside facing forward, the front fighting compartment housed the driver on the left and the radio operator and bow machine gunner on the right. Between them a rack held the radios, to their front the transmission, steering and brakes. Ammunition was stowed in the side panniers that overhung the tracks and in a compartment beneath the turret floor. The turret sat on a ring – central on the hull with three supports holding a turret floor. This floor in turn supported the hydraulic drive to turn the turret.

These five drawings were originally created for the wartime German *Tigerfibel* manual, but are presented here with annotations translated into English.

Cross-section through chassis.

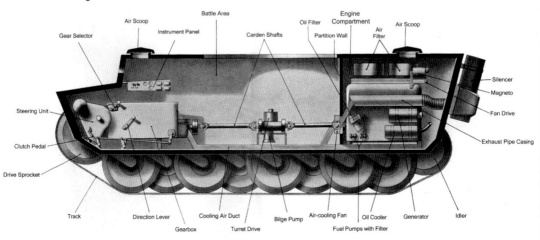

Battle Area

Engine Compartment

Air Scoop

Instrument Panel

Oil Filter

Air Filter

Air Scoop

Gear Selector

Carden Shafts

Partition Wall

Silencer

Magneto

Fan Drive

Steering Unit

Exhaust Pipe Casing

Clutch Pedal

Drive Sprocket

Track

Direction Lever

Cooling Air Duct

Bilge Pump

Air-cooling Fan

Oil Cooler

Generator

Idler

Gearbox

Turret Drive

Fuel Pumps with Filter

View from above.

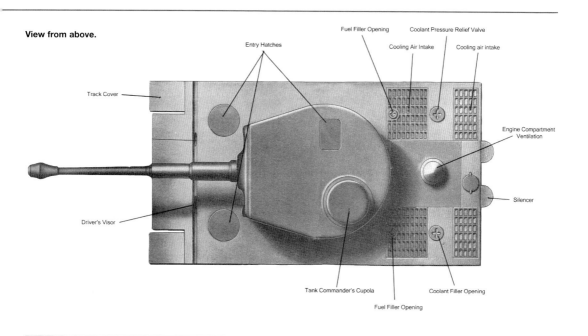

Track Cover

Driver's Visor

Entry Hatches

Fuel Filler Opening

Cooling Air Intake

Coolant Pressure Relief Valve

Cooling air intake

Engine Compartment Ventilation

Silencer

Tank Commander's Cupola

Coolant Filler Opening

Fuel Filler Opening

Top view.

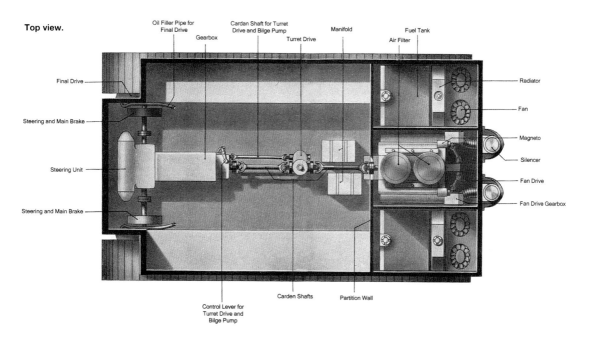

Oil Filler Pipe for Final Drive

Gearbox

Cardan Shaft for Turret Drive and Bilge Pump

Turret Drive

Manifold

Air Filter

Fuel Tank

Final Drive

Steering and Main Brake

Steering Unit

Steering and Main Brake

Radiator

Fan

Magneto

Silencer

Fan Drive

Fan Drive Gearbox

Control Lever for Turret Drive and Bilge Pump

Carden Shafts

Partition Wall

Tiger I – statistics (from Henschel, 1944)

Crew	5
Weights	
Combat weight	57,000kg
Rail weight	52,500kg
Speed	
Road	40kph (24.9mph)
Average terrain	20–25kph (12.4–15.5mph)
Range	
Road	195km (121 miles)
Average terrain	110km (68 miles)
Terrain capability	
Trench crossing	2.50m
Step climb	0.79m
Slope climb	35 degrees
Fording	1.60m
Submersion	4m
Length	
Gun forward	8450mm
Gun aft	8434mm
Hull only	6316mm
Overall width	3,705mm
Overall height	3,000mm
Width with cross-country track	3547mm
Width with transport track	3142mm
Ground clearance	470mm
Armour	
Driver's front plate	100mm at 9 degrees
Hull front lower	100mm at 25 degrees
Hull side lower	60mm at 0 degrees
Hull rear	80mm at 9 degrees
Deck	25mm at 90 degrees
Belly	25mm at 90 degrees
Engine	
Model	Maybach HL230 P45
Power	700hp
Cylinders	12
Weight	1,300kg
Turning radius	
Minimum	3.44m
Largest	165m
Fuel	
Capacity	540 litres
Fuel consumption per 100km	
Roads	270 litres
Average terrain	480 litres
Turret weight	11,000kg
Main armament	8.8cm *KwK* 36 (L/56)
Ammunition	92 rounds

ABOVE Looking down on the driver's hatch from above the mantlet.

BELOW The interior of the driver's hatch.

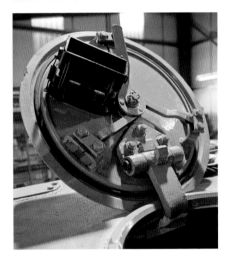

BELOW The driver's hatch closed. The stowed plate on the hull deck behind the hatch was used to seal an engine air intake during deep wading.

Looking around the vehicle

Hatches

The driver and radio operator both had circular hatches in the hull roof. These hatches had sprung hinges to ease opening and an armoured cover to house a periscope. When open a hasp held the hatch securely in position; no crew member wanted a heavy hatch to fall suddenly on head or fingers. The hatches had rubber seals and internal locking screws. It is impossible to drive the tank with a head out of the driver's hatch. When tanks are seen mobile in photos with a head or body of a crew member visible, it is another crew member sitting on the pannier step, on the hatch rim, or standing behind the actual driver, most likely giving steering advice from a better position with all-round visibility.

Driver's visor

The driver's only method of seeing forward to drive was through the shuttered visor on the front of the upright armour plate. This shutter met centrally when closed, controlled by a hand wheel to the right of the vision block inside the vehicle. The earliest Tigers were fitted with a pair of KFF 2 periscopes for the driver to see through if the shutter was closed. By February 1943, the periscopes were dispensed with and where front plates had already been drilled, the gaps were filled with weld. This can be seen on Tiger 131. The vision port was protected by a thick

TOP RIGHT The driver's position.

CENTRE RIGHT The Tiger driver's position looking forward. On the left is the *Kurskreisel* mounted on its bracket. Above the visor the plugged holes for the KFF 2 twin periscopes can be clearly seen.

BOTTOM RIGHT The driver's instrumentation panel. Spare laminated vision blocks for the front visor are stored above the panel. The connector box in the top left is part of a modern fire suppression system added during the original restoration.

RIGHT The *Kurskreisel,* or gyroscopic direction indicator, was fitted on the pannier surface to the left of the driver. Because tanks were made of ferrous metal they interfered with standard compasses. The *Kurskreisel* worked by having a powered gyroscope to ensure a stable axis. The upper dial, numbered from 1 to 12, could be set in the desired direction of travel by using the knob on top of the housing. The lower dial, also numbered 1 to 12, is attached to the gyroscope and therefore stays 'true', whichever direction the tank turns. The dial could be locked when the tank was parked and the power turned off. It could be calibrated using the side 'A' and 'E' switches. This is a mint boxed example.

RIGHT The face of the *Kurskreisel*.

RIGHT Diagram from a British intelligence report showing the arrangement of the turret front, the trunnions and the external armoured mantlet.

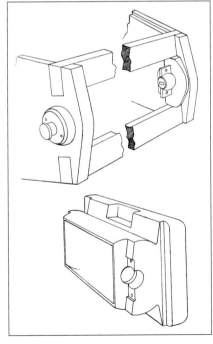

FAR RIGHT The cut-out section in the lower right corner of the mantlet indicates this turret is one of the 90 used that had originally been intended to fit the VK 45.01 (P).

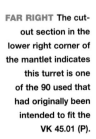

laminated glass block 70mm x 240mm x 94mm, spare blocks being kept next to the driver for ready replacement.

Mantlet

As part of the VK 45.01 (P) project, 100 mantlets were provided by Krupp. These mantlets had a small cut-out in the lower front right corner to allow the turret to turn and not catch on the raised engine deck of the Porsche-designed vehicle. The armour thickness around the sight was not considered

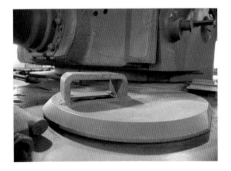

thick enough by the *Wa Pru* 6, so later castings had this rectified and the cut-out reinstated. Tiger 131 has one of the 90 earlier mantlets that were used (10 were not).

Commander's cupola

The commander's cupola or *Pz-Führerkuppel* is a drum-like structure with five vision ports and a hinged hatch for access to the turret. Each vision slit has a 90mm thick laminated glass

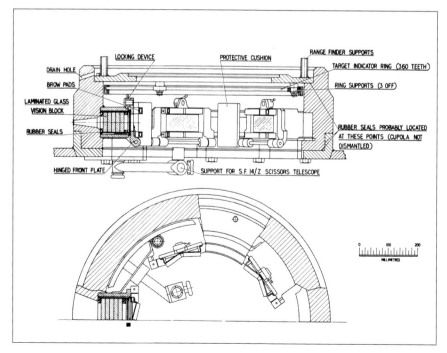

LOCKING DEVICE
PROTECTIVE CUSHION
RANGE FINDER SUPPORTS
TARGET INDICATOR RING (360 TEETH)
RING SUPPORTS (3 OFF)
DRAIN HOLE
BROW PADS
LAMINATED GLASS VISION BLOCK
RUBBER SEALS
RUBBER SEALS PROBABLY LOCATED AT THESE POINTS (CUPOLA NOT DISMANTLED)
HINGED FRONT PLATE
SUPPORT FOR S.F I4/Z SCISSORS TELESCOPE

MILLIMETRES

RIGHT Inside the tank, looking across the breech to the commander's and gunner's positions. The plaque on the turret wall (top left) lists the fourteen actions required to secure the vehicle for deep wading.

BELOW The loader's hatch in the ajar position, allowing air to flow into the turret.

BOTTOM The open loader's hatch could be secured in an upright position by a curved arm that fitted over a locating collar inside the turret roof.

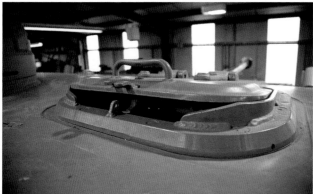

block behind that has a forehead and nose pad above and below it.

The cupola houses an azimuth indicator ring that is also repeated inside the turret beside the commander. The ring has 1–12 numbered around it and is connected to the turret ring. A similar indicator is located beside the gunner, allowing the commander to direct the gunner onto a target. The drum cupola was criticised by crews and later in the Tiger production, from July 1943, Tigers had a new cast commander's cupola with a much lower profile. It was fitted with seven periscopes and a domed hatch that raised and pivoted to the rear and left.

Loader's hatch

The oblong loader's hatch is hinged forward. Like the other hatches, it had a rubber seal for submersion. The central ring extends and locks the four retaining bars outward. The catch has a jaw allowing it to be closed in an ajar position to give the loader some air flow.

Ventilation

A fan was fitted to the turret roof to pull out the turret air (and remove fumes left from firing) at the rate of 12m^3 per minute. The external outlet was protected by an external armoured disk.

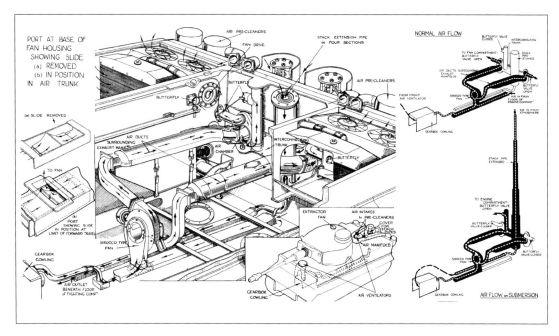

Labels in diagram (top-left): PORT AT BASE OF FAN HOUSING SHOWING SLIDE (a) REMOVED (b) IN POSITION IN AIR TRUNK; (a) SLIDE REMOVED; TO FAN; (b) PORT SHOWING SLIDE IN POSITION AT LIMIT OF FORWARD TRAVEL; GEAR BOX COWLING; SIROCCO TYPE FAN; AIR OUTLET BENEATH FLOOR of FIGHTING COMP.

Central diagram labels: AIR PRE-CLEANERS; FAN DRIVE; STACK EXTENSION PIPE IN FOUR SECTIONS; BUTTERFLY; AIR DUCTS SURROUNDING EXHAUST MANIFOLD; AIR CHAMBER; INTERCONNECTING TRUNK; BUTTERFLY; AIR PRE-CLEANERS; FROM FRONT AIR VENTILATOR; GEARBOX COWLING; EXTRACTOR FAN; AIR INTAKES to PRE-CLEANERS; COVER FOR STACK PIPES; AIR MANIFOLD; GEARBOX COWLING; AIR VENTILATORS; SIROCCO TYPE FAN.

Right diagram: NORMAL AIR FLOW; BUTTERFLY VALVE CLOSED; INTERCONNECTING TRUNK; TO FAN COMPARTMENT-BUTTERFLY VALVE OPEN; STACK PIPE STOWED; AIR DUCTS SURROUNDING EXHAUST MANIFOLD; BUTTERFLY VALVE OPEN; SIROCCO TYPE FAN; AIR IS DRAWN FROM FLOOR OF ENGINE COMPART'; STACK PIPE EXTENDED; AIR IN FROM ATMOSPHERE; TO ENGINE COMPARTMENT-BUTTERFLY VALVE OPEN; BUTTERFLY VALVE CLOSED; SIROCCO TYPE FAN; BUTTERFLY VALVE CLOSED; GEARBOX COWLING; AIR FLOW on SUBMERSION.

Vision slits

Two narrow vision slits were created in the front of the turret at the 10 and 2 o'clock positions. Each had a replaceable 90mm block of laminated glass fixed behind with a head and nose pad.

Emergency hatch

The *Notausstiegluke* (escape hatch) was fitted at the rear right of the turret. Locked from the inside, it was hinged at the bottom and simply fell back when opened. The weight of the hatch caused complaint as it was often used to assist in reloading the vehicle and at times a port from which to fire small arms. The weight of the hatch meant it could not be closed from inside the vehicle; someone had to stand on the engine deck to lift the hatch back into place. Despite complaints, the design was not altered during the production run.

Baggage bin

Very early production Tigers were provided with the stowage bins of the type fitted to the Panzer III. Later tanks were issued with a much larger *Gepäckkasten* (baggage bin) that was shaped to the rear of the horseshoe turret and had two hinged lids.

The third lifting stud for the turret is at the six o'clock position on the turret rear. In consequence, a cut-out is shaped into the bin.

The metal bin was used to store spare track links and pins, tools and crew baggage such as personal packs, blankets and canvas. Badly stowed material, if left on the outside of the vehicle, could interfere with the rotation of the turret or cause a fire risk if allowed to touch a hot exhaust or engine part. The bin on Tiger 131 has a padlock left in place by the crew that is still affixed. The padlock – never broken off as it was locked in place with the bin open – can be seen in early shots of the tank immediately after capture. Despite the tank being taken

Water Tiger – submersion and wading

A nother feature the designers of the Tiger had to address was the requirement for submerged fording of water obstacles. Concerned with the weight of the Tiger the German authorities stipulated the design must incorporate a deep-wading capacity. Rather than find bridges with enough capacity to take the weight of the vehicle, it would cross rivers and water obstacles to the depth of 4.5m by the use of a *snorkel* system.

This requirement led to a number of special provisions on the Tiger that created much extra work during production and there is no evidence that the feature was ever used on campaign. The realisation that this was an unnecessary feature led *Wa Pruf* 6 to cease the production of the submersible element of Tigers in August 1943. Instead, they ordered the Tiger to be able to ford to a depth of 1.5m.

For a Tiger to deep wade the crew had a series of preparations to carry out – fourteen are listed. If any one of these was missed it might lead to the flooding of the vehicle. As a checklist, a plate was fixed to the turret wall listing all the actions required and a corresponding number painted on the relevant piece of the tank. The checklist plate is missing from Tiger 131 but photographs show it in place after capture. The procedure meant all hatches and openings had rubber seals or plugs, the turret ring an inflatable rubber tube to stop any water ingress and a large threaded standpipe was erected on the engine deck to provide air to the crew and engine.

A fan drew air beneath the turret floor and back through the exhaust manifolds and exhausted through butterfly valves in the engine compartment. The *snorkel* pipe was made of four pieces that telescoped inside each lower section and when not used, it was housed in the engine bay, upright on the inner rear wall.

As part of the submerging process, the fan drives were disconnected and water flooding in around the radiators provided cooling. A bilge pump set under the turret floor

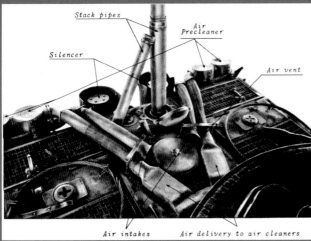

Stack pipes

Air Precleaner

Silencer

Air vent

Air intakes

Air delivery to air cleaners

ABOVE A view of our Tiger's engine deck, copied from an official report, showing the air cleaner system and telescopic breathing tube partly erected.

RIGHT After the war a Tiger was used at the Haustenbeck trial. The unique location for the *snorkel* on this tank can be seen on the commander's cupola.

LEFT Tiger trial crew and tank.

FAR LEFT AND LEFT The body of the tank is fully immersed.

BELOW The Tiger emerges from the water tank. The domed MG34 cover can be clearly seen on the front plate.

scavenged any leakages. The bilge pump could remove 250 litres of water a minute from the vehicle through a pipe to the right side of one of the fuel tanks. It had a spring covering to stop any water flow back into the vehicle. The pump needed anti-freeze to stop it freezing up in sub-zero conditions.

At the end of the war, a Tiger was evaluated at the German testing site at Haustenbeck by a British team interested in gaining any useful information on the submerging capabilities of the tank. For deep-wading experiments in a special tank at the site, the Tiger had the *snorkel* fitted to the commander's cupola, not in the usual position, at the rear of the engine deck.

Fire extinguishers

The Tiger had a fire suppression system fitted in the engine bay. Temperatures over 120 degree C would trigger a 7-second burst from a 3-litre extinguisher directed at the fuel pumps and carburettors. The extinguisher would continue with up to five 7-second bursts in trying to eliminate the heat source. Within the turret, a hand-held fire extinguisher was fitted to the turret floor for ready access.

MP port

Very early Tiger turrets had two *MP-Klappe* or machine pistol ports in the turret rear to allow the use of an MP 40 or pistol at close-range targets. The port was opened by rotating an internal lever. From turret 46, the second port was replaced by the emergency escape hatch on the rear right quarter. Part of the *Tigerfibel* illustrated the 'blind spots' surrounding the vehicle wherein enemy troops might try to approach. *S-Minenwerfer* (S-mine dischargers) were fitted to some vehicles to help protect against tank hunting teams approaching the vehicle. Later during the production run a *Nahverteidigungswaffe* (close-defence weapon) was fitted in the turret roof. This small projector was angled at 50 degrees and could be rotated 360 degrees and fire a range of projectiles such as smoke and anti-personnel charges to detonate near the vehicle. These systems were of course just as deadly to friendly troops as they were to any approaching enemy.

Feifel air filters

Two pairs of filters or pre-cleaners were fixed to the rear corners of the tank, taking air through tubes that rested on the engine decks and back through similar tubes into the engine bay. The system was dropped in October 1942.

apart for evaluation and over years of study, with people crawling on it and now running, the padlock has remained steadfastly in place. The bin also shows evidence of battle damage.

Turret lighting

The Tiger had two 12-volt 150 amp/hour batteries charged from a generator when the main engine was running. The importance of the batteries to start the engine, run systems such as the radio and illuminate the tank, led to the installation of a battery heater on tanks manufactured after December 1943. The batteries lit the instrument panel for the driver, a light for the radio operator and three turret lights. These were simple metal tube arrangements with a cover that could be twisted open to regulate how much light was revealed.

Wheels and tracks

The wheels that support a tank on the ground are there to spread the weight of the heavy vehicle across the tracks. If there were no tracks, the vehicle would simply sink into the ground. The drive wheel at the front of the Tiger locks into the track with the toothed sprocket wheel. This in turn is powered by the engine and pulls the tank along, laying track in front of the vehicle in a continuous band. The rear idler

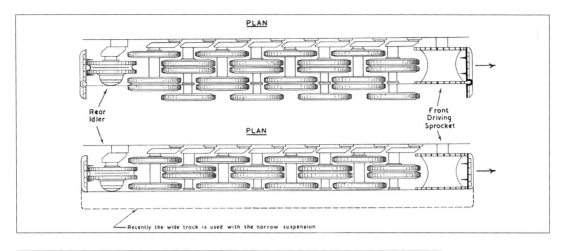

PLAN

Rear Idler

PLAN

Front Driving Sprocket

Recently the wide track is used with the narrow suspension

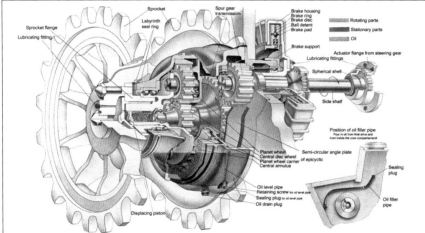

Sprocket flange
Lubricating fitting
Sprocket
Labyrinth seal ring
Spur gear transmission
Brake housing
Brake ring
Brake disc
Ball detent
Brake pad
Brake support

Rotating parts
Stationary parts
Oil

Actuator flange from steering gear
Lubricating fittings
Spherical shell
Side shaft

Position of oil filler pipe
Pour in oil from final drive and from inside the crew compartement

Planet wheel
Central disc wheel
Planet wheel carrier
Central annulus

Semi-circular angle plate of epicyclic

Oil level pipe
Retaining screw for oil level pipe
Sealing plug for oil level pipe
Oil drain plug

Displacing piston

Sealing plug

Oil filler pipe

ABOVE Layout of the road wheels and suspension arms showing the difference between the *Marschkette* (operational track suspension) and the *Verladekette* (transport track arrangement).

LEFT Final drive and double-drive sprocket.

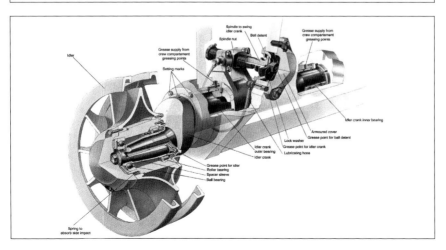

Spindle to swing idler crank
Spindle nut
Ball detent
Grease supply from crew compartement greasing points

Idler
Grease supply from crew compartement greasing points
Setting marks

Idler crank inner bearing

Armoured cover
Grease point for ball detent
Lock washer
Grease point for idler crank
Lubricating hose

Idler crank outer bearing
Idler crank

Grease point for idler
Roller bearing
Spacer sleeve
Ball bearing

Spring to absorb side impact

LEFT Idler with track tensioner.

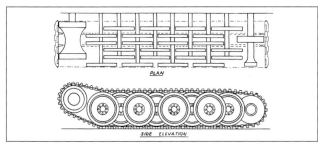

ABOVE A simplified diagram showing the overlapping and interleaved road wheel arrangement on the left side of a Tiger viewed from above. Note that on the opposite side the road wheels are mounted on leading arms, not trailing arms as they would be here. This is typical of armoured vehicles with lateral torsion bar suspension and means that wheel stations on opposite sides of the tank are not directly in line.

RIGHT The width of the interleaved suspension can be clearly seen in this unusual shot of the belly of an upended Tiger in the factory. In the top left of the hull floor the water drain plug hole can just be made out. More central and larger is the round transmission oil drain point. In the lower left of the hull floor can be seen the round access point to drain engine oil and water.

RIGHT The lateral torsion bars are exposed during the restoration of Tiger 131. The raised plates hold grease nipples that connect to and lubricate each swing-arm housing.

BELOW Fitting the torsion bars at one of the *Takt* (or production) stations. The bar that is seen being fitted will cross the hull floor and have a swing arm attached on the far side of the vehicle.

BELOW This is one of the connectors that attaches to the torsion bars. The inner collar of the larger socket is grooved to meet the machined end arm of the torsion bar.

wheel keeps the track central and of a correct tension so it does not slip off.

The Tiger's weight led Henschel to adopt an interleaved suspension system or *Schachtellaufwerk*. By this method, 24 wheels were fitted to each side of the hull in groups of three. This spread of road wheels was necessary as the rubber-rimmed wheels failed with too great a load, at too regular an interval, needing replacement. Later when metal-rimmed wheels were used to save on rubber, the Tiger's outer layer of wheels could be dispensed with completely.

The wheels were attached to a torsion bar suspension system. In this system a metal rod is secured on one side of the inner hull and crosses to the opposite side, exits through a drilled hole and is attached to a suspension arm. In turn, the arm is attached to three road wheels and as the arm is forced upwards as the wheels travel over an obstacle, the bar is twisted. The metal's natural springing action returns the suspension arm back downwards. The front and rear torsion arms were also fitted with hydraulic shock absorbers.

The width of the Tiger was 3,547mm wide with the standard *Marschkette* (or operational) tracks fitted. This meant the Tiger would not fit the Berne loading gauge. The design team met this problem by allowing the outer road wheel to be removed and a thinner set of *Verladekette* (or transport tracks) to be fitted. Very early production Tigers had no mudguards fitted to the hull sides, but as the *Marschkette* tracks were wider than the hull,

ABOVE With the *Verladekette* track fitted. The outer layer of road wheels and track guards have been removed and the front hinged guard folded back.

BELOW This newly completed Tiger is issued on its transport tracks.

LEFT Tigers of the 501st Heavy Tank Battalion are loaded onto ferries for the sea crossing to Tunisia. No Tigers were lost in this or later shipments to North Africa.

LEFT Once the tank was aboard the railway flat it was chocked and anchored in place. Here a Tiger of the 501st is prepared for transportation.

they could throw considerable amounts of mud up onto the hull deck. Mudguards were then fitted and they too had to be removable. The front covers had hinged wings to also allow them to fold in and meet the width requirements.

To test whether the ground could take the weight of a Tiger, the *Tigerfibel* suggests carrying a crewman on piggy back and standing on one leg to see if the ground is hard enough to support you both. If your foot starts to sink – don't drive there.

On the Russian front, with few metalled roads, improvised corduroy roads made of logs were built. If a Tiger was to traverse one, the logs had to be at least 15cm in diameter and 3.5m wide or the tank would rip them up.

BELOW These vehicles have the full-width *Marschkette* (operational tracks) fitted.

ABOVE Here is an example of a sectioned steel-rimmed road wheel. The rubber is the small black domed section under the central circular ridge. It was sandwiched and held in place between the outer rim and inner wheel section to act as a cushion by the circle of bolts. Belatedly the German authorities pursued the compatibility of parts, with this wheel type fitting the Tiger I and II and Panther series.

New road wheels

In February 1944 the Tiger underwent a visibly distinctive change with the introduction of rubber-saving road wheels. The new *Gummigefederten Stahlaufrollen* – or rubber-cushioned steel-tyred road wheels – only had rubber as an internal ring bolted between the inner wheel disc and the outer solid steel wheel rim. The new wheels led to the elimination of the outer layer of wheels, making the pattern now a single inner wheel, a double wheel and then a single outer wheel.

ABOVE Various parts from Tigers that were blown up at Hunt's Gap in Tunisia. Seven Tigers were destroyed at the location – five ran over mines, one became stuck and the other was hit by artillery fire. The left wheel in this picture has been marked for transit to the 501st and the weight for transit, 77kg, has been painted on it.

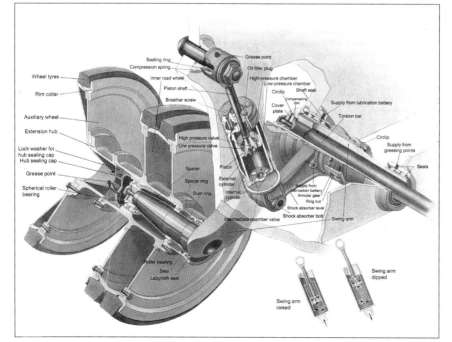

LEFT Wheel shock absorber.

Armour

When the Tiger rolled into action, the crew had not just the reassurance that the firepower of the tank could outrange their opponents – they also had the comfort that the armour protection was initially impervious to the tanks they were likely to meet. This in turn, added to the mobility of the tank on the battlefield where often the tank could manoeuvre in a manner other vehicles dare not.

The Tiger's armour can hardly be called a stylish or sophisticated affair (if indeed these terms can be used about any armour protection on a Second World War vehicle). The protection given to the tank was provided by the thickness of the material used, presented at almost right angles to the incoming rounds.

The armour was primarily made up of cut shapes of rolled homogenous armour plate that was overlapped, stepped or dovetailed and then welded together. Rolled Homogenous Armour (RHA) was of the same consistency throughout, not face-hardened as some armour. The rolling process increased the hardness of the material but with the Tiger it was the thickness of the material and the quality of the joints that gave protection. The specifications for the armour provided by the manufacturers had subtle differences in alloy composition. *Wa Pruf* 6 standardised the alloy content during the production run of the Tiger but dependent on the thickness of the plate to be manufactured, a differing mix of alloys were used.

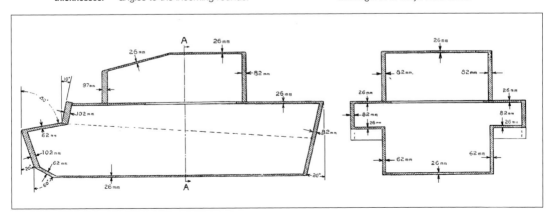

The frontal armour on the driver's plate was 100mm thick, set back at a gentle 10-degree angle from the vertical. Below this the sloping glacis plate was 6cm thick at 80 degrees from the vertical and the frontal armour 100mm thick sloping back 20 degrees from the vertical. The deck plates were 25mm thick and the vertical side armour 60mm. The rear plate, sloping back at 9 degrees from the vertical was 80mm and the belly plate 25mm thick. The hull of the Tiger was welded together from the component plates and the top superstructure then bolted to the sides and welded in place.

The turret armour consisted of one large piece of armour 80mm thick bent to horseshoe shape. Straight pieces welded across the front of the horseshoe provided the frame for the mantlet. The roof plate was in two parts and was 25mm thick over most of the area, a second inserted piece of metal at the front of the turret roof was 40mm thick.

The dovetailing and interlocking of the plates was designed to give further structural and protective strength. The crews were taught to increase this armour protection by judicious positioning of the tank in combat.

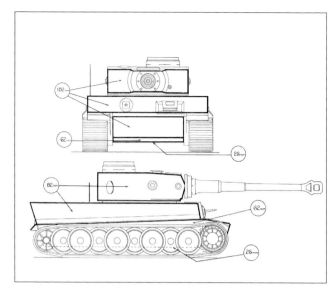

The *Tigerfibel* explains how by using a clock system (with 12 o'clock being with the gun facing straight ahead) the tank should be positioned to the enemy at 10.30, 1.30, 4.30 or 7.30 positions. By doing so the angle of the armour creates a thickening effect to the incoming round.

ABOVE Armour thickness on the Tiger in millimetres.

Unteroffizier Doctor Franz-Wilhelm Lochmann, Schwere Panzer Abteilung 503

'An example of a long-range engagement, say 2,000m or so, there was a whole line of T-34s moving along a road at right-angles to us in the mountains in Czechoslovakia. Then our commander said to the gunner 'hit the leader and then the last one'. Then you had time and could get them all, one after another. The crews could see they were helpless and all bailed out. They were side-on too. That's bad.

'For us to be knocked out, they had to put themselves in a position where they could shoot at us from the side into the lower hull. It was important to us to use the angles to help us – the tanks were like barn doors so our tactic was never to expose our sides to the enemy, but always stay slightly at an angle.'

LEFT This destroyed Tiger turret clearly reveals the bent horseshoe-shaped armour that made up the turret sides.

Zimmerit – Mike Gibb

Zimmerit was a special paste that was used to coat German armoured fighting vehicles during the Second World War for the purpose of combating magnetically attached anti-tank mines. Tigers had *Zimmerit* applied starting in August 1943 at the factory and units in the field were ordered to apply the coating in January 1944 – 200kg of the material being needed to complete a tank to the correct thickness. Ironically, none of the Allied nations used magnetic mines, so all the energies put into the coating process were wasted.

The following description by Mike Gibb shows how *Zimmerit* was remanufactured so it could be applied to the *Sturmgeschütz* that

RIGHT Ingredients for *Zimmerit*: zinc sulphide, barium sulphate, pine saw dust, PVA, pebble dust, ochre and pine crystals dissolved in benzene.

RIGHT Trowelling on the material to the hull side.

the SdKfz Military Vehicle Foundation restored in 2008:

Zimmerit, the anti-magnetic mine paste, was applied on all StuGs leaving the assembly plants between August 1943 through to September 1944. The concept of the Zimmerit AG product was that the paste would provide a protective layer on the vehicle against magnetic mines. The compound's application was done in such a way so as to break up the surface into a series of small ridges or steps producing a highly uneven surface layer. The paste was not anti-magnetic but provided an uneven insulation layer of 6mm which was supposed to reduce the magnet's hold on the steel surface beneath.

To our knowledge for the first time in 65 years, we reproduced *Zimmerit* to the original detailed formula and applied it to a vehicle. The correct recipe, including ingredients and weights/cubic volumes, were gleaned from a variety of sources including a British Intelligence report, which provided a chemical analysis of what made up *Zimmerit*. After the war, interviews were held with personnel who were responsible for the *Zimmerit* programme, members of Zimmerit AG and shop floor supervisors. The original application instructions were also included.

It is interesting to note that the chemical analysis provided an accurate comparison to the wartime *Zimmerit* records on weights and cubic densities that emerged after the defeat of Germany.

Instructions for the application of *Zimmerit* in the StuG plants between August 1943 and September 1944 was for 75kg of the paste to be applied. Specific instructions about where to apply the material, and where not to, still exist. These instructions were generally well adhered to but the firms of ALKETT (Altmärkische Kettenfabrik) and MIAG (Mühlenbau und Industrie) went their own way on the method of application. Instructions on how to apply the 75kg of *Zimmerit* were very specific – ridges applied by a spatula or trowel uniformly applied over the designated surfaces. ALKETT went on to characterise their vehicles with a distinctive waffle pattern while MIAG elected to cover their product with 50mm hand-drawn squares. Why both assembly firms chose a differing design

is unknown, but this now provides a quick and easy way of distinguishing between MIAG and ALKETT vehicles in the period of August 1943 and September 1944.

A most gratifying by-product of our real-time experiment in reproducing *Zimmerit* was the dispelling of a number of myths surrounding the material. The ingredients of the paste: zinc sulphide, barium sulphate, pine saw dust, PVA, pebble dust, ochre and pine crystals dissolved in benzene had caused some concern as *Zimmerit*, according to many after-action reports, had reputedly caused fires to break out without the armour having been pierced. This assertion had been rejected by the authorities who conducted a test of their own by firing numerous shells against a *Zimmerit*-coated vehicle with the inevitable result – no fire risk.

Our own experiment proved the opposite. Dissolving pine crystals in benzene requires a considerable amount of benzene. This creates an extremely sticky, clear golden-coloured liquid which, when added to the other components, assists PVA in providing the adhesive and hardening qualities in the finished product. The instructions require the use of a blow lamp held 5cm away from the surface to harden and burn off excess moisture after the first application of 2mm and then the last 4mm. This is only to be done four hours after each application – allowing the *Zimmerit* time to dry. The fires that result from the blow lamp burn-off on the last application are significant. The surface is then painted but will not harden in a cold environment. A cold environment could almost be guaranteed in the winters of 1943–44, not only because of mother nature, but also due to the efforts of the US 8th Air Force and RAF

Bomber Command. This meant that a vehicle would leave the assembly plant with an, as yet, un-dried *Zimmerit* exterior. The paint was locked into the benzene, which had not evaporated, and the armoured vehicle was delivered into harm's way – the battlefield. Who was telling the truth? The men on the ground or the authorities?

ABOVE LEFT AND RIGHT Burning off and hardening the surface as instructed could cause interesting results.

ABOVE *Zimmerit on the completed MIAG-made SdKfz142/1 Sturmgeschutz restored by the SdKfz Military Vehicle Foundation.*

LEFT A *Zimmerit-covered Tiger, knocked out in Normandy in 1944.*

'The Tiger is not the lumbering beast portrayed in many books and films; it has the same mobility and ground pressure as the Panzer IV; the vehicle is very agile and when driven correctly it is quite fast for its size and age.'

Stevan Vase, Tank Museum workshop volunteer

Chapter Four

Restore to Running Order

David Willey

With Tiger 131 delivered to the Tank Museum at Bovington in Dorset, a new era began for the vehicle with a new set of issues. This chapter reveals how the Tank Museum has met the challenge of how best to preserve, display and interpret a modern mechanical item.

OPPOSITE Running historic machinery can add considerably to the understanding of the subject matter and it undoubtedly brings people to the museum, as this photograph taken at the Tankfest event in 2009 shows.

The reason why

Tiger 131 was handed over to the Tank Museum with much other captured equipment in 1951 from the Ministry of Supply. It was entered into the accession books as Accession 2351 (now E1951.23). It had been the subject of thorough analysis at Chobham and had a considerable amount of work carried out on it during the evaluation to discover information and, while needed, to keep it in running order. At a later date, the Maybach HL210 engine was removed from the vehicle. As was standard for much captured equipment that was to be used for instructional purposes the engine was later sectioned to help in its use as a teaching aid for serving soldiers.

Some of the activities carried out by the Army in the disassembly and analysis of captured vehicles have caused groans from subsequent generations of tank enthusiasts. Items were cut from tanks and sent for metal analysis or destructive testing. However this process was considered a necessary part of gathering intelligence and information on enemy equipment – quite simply a war had to be won and the heritage value of such an item would understandably be of little importance at this time.

ABOVE The Tiger was gifted from the Ministry of Supply to the Tank Museum collection in 1951. In the accession book it is entry number 2351, 'German TIGER. (E)'.

RIGHT Part of the line-up of German vehicles in the Tank Museum, probably during the late 1950s. The Tiger now has large outlined turret numbers.

What is original in a tank?

With the Tiger now at the Tank Museum, though the same machine, a new set of issues and priorities arise related to the tank as it is now a collection item. One key area is originality, an issue that has relevance to all museum collection items but has a particular complexity with mass-produced machinery such as a tank.

While the Tiger is clearly a historically significant item in both general terms and with the known battle history of this particular vehicle, the museological question of originality is far less easy an issue to define.

The originality (or the authenticity) of an object and the evidence and information this supplies is often seen as of primary importance in a museum item. Think of a work of art and the day it leaves the artist's studio. This might be viewed as *the* moment of completeness reflecting the artist's intention and therefore the originality or authenticity a museum most wishes to retain. Subsequent years of movement, the addition of dirt, cleaning, re-touching, re-framing, in fact any alterations, however interesting the reasoning or reflective of changing times, only takes away from the artist's intention.

But with a mechanical item how do we define originality – and who makes the decision? Is a tank, for example, original the day it leaves the factory? Or after soldiers adapt it in service? After it receives battle damage, a new gun or replacement engine? Did the Tiger end its service life with the *Wehrmacht* on 21 April 1943 or later at Chertsey under British analysis? Or docs it still have a service role as it is pointed out to countless serving soldiers on their tour of the museum learning about tank development?

These questions can understandably cause endless debate and there may be no real right or wrong answers, just better argued cases.

The Tiger at the Tank Museum

The Tank Museum has a teaching and reference role for the Royal Armoured Corps Centre at Bovington (that continues to this day) so early forms of interpretation of the vehicle were directed towards the serving soldier – tanks were lined up by country to show national developments. Though the tank looks complete, staff recall that the vehicle was not actually a fully reassembled item, with parts missing and other components stored separately. A number was painted on the front of the tank to relate to a catalogue and at some stage a spurious paint scheme and markings were added. Steps were attached to the side to allow access and, as with many vehicle collections, the standards of care differed considerably to those of today.

The Tank Museum and operating vehicles

The Tank Museum was often involved in the Royal Armoured Corps open days or 'Battle Days', where the Army demonstrated serving military vehicles to the public. Newsreel footage and programmes show that a historic element was almost always in evidence with vehicles that could be run from the museum taking part. The popularity of these displays and the increasing need to attract a paying visitor audience led to more regular displays of running vehicles in the 1990s. These displays tended to use vehicles the museum could acquire or had multiples of, gifted at the end of the Cold War – Chieftain tanks and Ferret scout cars for example. But the interest

BELOW Royal Armoured Corps Centre open day souvenir book covers.

in the rarer items remained and as Bovington was recognised as 'the home of the tank' and the museum's collection was considered to be the best in the world, the emphasis grew on restoring more key vehicles to running order.

As with many museums that run collection items, the Science Museum being a good case in point, policies as to what could be run and how tended to be developed on an ad hoc basis. The Tank Museum was no exception; policies and procedures only gradually became established – the most important was the formation of an identified running fleet of vehicles for regular display use and the gradual accumulation of vehicles for this fleet. This would lessen the impact of regularly running vehicles from the core (or unique vehicle) collection.

The museum decided to continue a policy of investigating vehicles as potential runners on a case by case basis – if items were considered time capsules, or return to running order likely to require considerable intrusive work, the decision was made not to run. Some vehicles that had been returned to running order in the past were retired from running as the risks to the historic integrity of the vehicle by further running was considered too great.

Why run vehicles?

The incredible value that running vehicles can provide for the interpretation of the collection can be seen in the popularity of our 'Tanks in Action' displays where running fleet vehicles take part and the now annual 'Tankfest' where selective core collection items are taken out to drive. It has reinforced the museum's position that running vehicles is a core activity in interpreting the collection to the public and this was acknowledged by the building of a £1m arena for viewing as part of the museum's recent £16m Heritage Lottery Fund (HLF) development.

For museums the issues surrounding the running of collection material is of course very complex. Running unquestionably compromises originality and can lead to the loss of certain historic evidence. However, it can also reveal new information about the collection item, sustain skills in operation and maintenance, and benefit the interpretation, understanding and overall value of the collection to the public and considerably help the overall sustainability of the museum; a practical but important aspect often overlooked in theoretical debate on collections.

The Tank Museum has been pragmatic in its approach to running vehicles and believes that each case should be taken on its merits. 'One size fits all' may not be the best approach to the issues – a view that has in the past polarised opinions in the museum and heritage communities. The issue will always cause controversy among certain enthusiasts who would wish to see everything to run, to the 'over my dead body' school who would rather there was no potential for wear or subsequent loss of evidence. Current trends seem to indicate a much broader church consensus emerging rather than the 'us and them' attitude that many fear or assume.

The decision to restore and run the Tiger

To return to the late 1980s, the Tiger was considered such an iconic vehicle, despite its missing parts and overall complexity, that it became a Tank Museum goal to return the vehicle to running order. For Tiger 131 the decision was taken to try and return the vehicle to as near to its state at capture as possible, with, as might be expected, certain compromises. Damage that had been repaired in North Africa or at Chobham was left in place. The engine would always prove a problematic issue and it hurt some veterans to see the shield of 1st Army and the diabolo of 21st Army Tank Brigade removed from the front hull where they had been painted by proud new owners.

Standards and objectives have of course changed over the intervening period but it would be interesting to pose the question, would the Tank Museum come to the same conclusion to run the tank today as it did in the early 1990s? Curator David Willey says, 'My own feeling is probably yes – we would like to return the vehicle to running order, but we would now do it in a different way.'

The restoration of Tiger 131

Initially, the project was carried out by staff within the museum with modest grants made by the Winston Churchill Memorial Trust to research the history of the tank. A number of defence companies and related industries came forward to offer assistance. Despite much progress over a number of years, the time and complexity of some of the issues involved in reassembling the vehicle, the loss of workshop space and overall resource issues, led the Director of the museum to stop the in-house restoration and assemble a bid to present to the Heritage Lottery Fund to fund and commission an external restoration.

The bid for funds to restore the Tiger was made together with bids to restore a First World War vehicle – the Whippet tank. After due consideration, the Heritage Lottery Fund decided against funding the Whippet as it was considered too much a historical time capsule whose completeness and vulnerability would be compromised if returned to running order. However, the Tiger with its history of disassembly and work already begun, was approved. A sum of £96,000 was awarded by the HLF towards a project cost of £184,000.

With hindsight the use of the Army's Base Repair Organisation (ABRO – at the time a 51 per cent government-owned agency that repaired and refurbished the Army's vehicle fleet, but which also took on outside contractual work) may seem an odd choice for the restoration of a historic item like the Tiger. However, in the mid-1990s, few private individuals were restoring German armoured vehicles, so the pool of knowledge in the field was very limited. No one had returned a heavy German tank to running order. ABRO offered the Tank Museum genuine experience and skills in the movement and refurbishment of complex heavy armoured vehicles. They were also on the Bovington Camp and had considerable status and backing – an important issue when dealing with the HLF who, as an organisation, was understandably risk averse when agreeing to contractors.

The restoration of the vehicle by ABRO threw up many issues – not all of which were successfully addressed at the time. The control

of the restoration process off-site was a new challenge for the Tank Museum and, with hindsight, not enough oversight was given to the restoration. In 2000, a number of key issues emerged – one of which concerned the engine.

The use of the historically accurate but later model engine – the Maybach HL230 in place of the sectioned HL210 – had been agreed early in the restoration project. The museum had a number of Maybach engines potentially available, but no complete HL210. The use of the HL230 was started by the Germans in their Tiger Is in May 1943 and earlier model tanks (such as Tiger 131) were issued with field conversion kits to allow the later model engine to be fitted when the earlier model engine needed replacement. The conversion included new fan drive gearboxes, as

ABOVE The aluminium-blocked Maybach HL210 that was sectioned as a teaching and display example. The engine is resting on its side to reveal the cylinders and tappets.

BELOW These are the fan drives that take power from the engine and transfer the drive through the engine bay wall to turn the radiator fans.

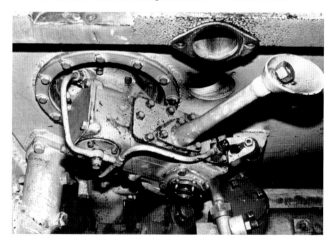

ABOVE The sets of
fans before fitting.

soldier-proof – the Maybach HL series was at
the cutting edge of engine technology in 1942.
The delicate assembly and disassembly by
ABRO was a process that had to be learned
from scratch – little existed in the way of
contemporary literature and no other Maybachs
were in working order in the UK.

Running the engine in the hull of the vehicle
during a test drive led to a repeat of a known
wartime engine failure – oil starvation leading to
the breaking of a con-rod and cracking of the
block. This occurred during the build-up to the
unveiling of the tank in July 2001. The problem
was solved by the repair of the block and in
turn showed up the problems and benefits
of the use of ABRO as restoration agents.
While they had no prior experience of work
on such historic engines, the resources of the
organisation could accommodate extending the
restoration project. The then Chief Executive of
ABRO guaranteed the return of the vehicle to
running order and the organisation continued
working on the project well beyond the
estimated time and paid hours agreed.

The need to complete the job after a lengthy
overrun led to compromises in a number of areas.
As with all major restorations, new problems were
discovered during the restoration process. The
HLF's appointed monitor, Tom Wright (ex-Science
Museum), was an excellent source of advice
and guidance during this process. The tank was

the power take-off to drive the cooling fans came
from the sides of the new model engine, not the
rear as on the earlier model.

Only during the restoration process did the
importance of these missing fan drives become
fully apparent, and after lengthy searches and
discussion with the HLF monitor a temporary
modern system was agreed on and made by
Supacat – a local engineering company. This
improvisation was considered essential at the
time to meet the HLF contract and return the
tank to running order.

Another issue that caused concern was
the rebuilding of the HL230 engine. Unlike
most modern tank engines that are robust and

RIGHT The Maybach
HL230 is run up on
a test stand before
refitting to the vehicle.

FAR RIGHT The tight
fit of the engine
inside the bay can
clearly be seen as the
Maybach is dropped
into place.

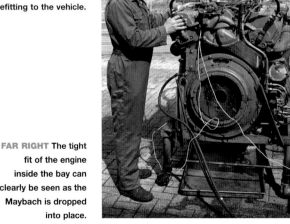

LEFT A test-track run for the engine in the hull. A connecting rod has just smashed through the engine block.

driven to meet the terms of the HLF contract but all agreed further work would be required. The engine ran hot and had oil pressure issues on its first outing at Tankfest 2004.

A chance for a second look at the engine was provided when the SdKfz Military Vehicle Foundation offered assistance, in exchange for surplus parts in the restoration of one of their vehicles. Mike Gibb of the Foundation had to face problems similar to the museum's in returning a Maybach HL230 engine to running order. This was for a *Jadgpanther* the organisation had restored. The knowledge, skills and engineering facilities the Foundation had built up in the restoration of German vehicles was second

ABOVE The block after the engine failure. The white card indicates where the hole was punched through the casing.

LEFT The ABRO team: Ron Luke, Nick Morgan and Dave Marsh with Tom Wright (the Heritage Lottery Fund-appointed monitor) in front of the tank on the day it ran to complete the HLF grant conditions.

ABOVE The SdKfz Military Vehicle Foundation's *Jagdpanther* SdKfz 173 at Tankfest. This camouflage scheme was introduced during the Second World War as a paint-saving measure. A third of the red oxide primer was left uncovered.

to none, and co-operation between the Tank Museum and the Foundation developed.

With the Tank Museum's oversight, the Tiger's Maybach engine was stripped and rebuilt by the SdKfz Military Vehicle Foundation, then returned to the tank to give a much smoother-running vehicle.

The continuous need for work

The Tiger has been run at events for the public and filmed and recorded for a number of years. The vehicle is the centre of tremendous interest (tap in 'Tiger tank' on YouTube and see how Tiger 131 appears). Its movements, running time, oil consumption and maintenance carried out is carefully monitored and recorded.

In many ways the restoration of such an item

will never be complete. More original components may come to light, and a number of areas that were not restored or reinstated in the original HLF-sponsored restoration can be addressed. For example, the inertia starter, the original fuel tank, main engine water pipes and internal fittings are yet to be installed. This will help bring the vehicle back to a state of completeness representative of the condition at its capture, and create a much more authentic and complete vehicle. The temporary fan drive arrangement was just that, a temporary affair and the reproduction of new fan drive gears to original plans will bring the vehicle back to an even more original state.

Running the vehicle will have obviously created wear and so the museum is using the opportunity of further work to investigate the wear rates and the wider effects of operating such a historic vehicle.

How much, or whether, the vehicle continues to run into the future at all is a matter of continuous assessment. The museum wants to analyse the effects of running the Tiger and disseminate this information. Knowing the appeal of the vehicle and the huge interest in seeing the tank run, the museum would like it to continue running – but certainly not at any cost. If wear, breakages and risks are considered too great, the tank will stay static. If wear rates and risks are considered acceptable, the museum will hope to run the vehicle for set periods before another major investigation.

RIGHT The tank displayed part-way through the restoration. Although the subject of much interest, the restoration of large mechanical objects is not a continuous process, punctuated with hitches that so many television documentaries portray. Restorations take a long time, in a low gear and have many inevitable periods where nothing much happens, as components are sent away or parts are made.

Restoration

The following images show some of the stages in the restoration process of Tiger 131.

1 Removal of the Tiger's gearbox and steering unit.

2 The narrow fit of the gearbox and steering unit through the turret ring can be clearly seen in this image. On the right, Harry Webb, a Second World War tank veteran, helps his old adversary.

3 The hull is taken away to an empty vehicle workshop on the Bovington site to be worked on.

4 The tracks are removed and stored.

5 Removal of the road wheels. Here, battle damage can be clearly seen.

6 The turret is placed on a stand and worked on separately. The weathered appearance makes the paint look original, but together with the numbers it is in fact a later addition.

7 Removal of the gun mantlet.

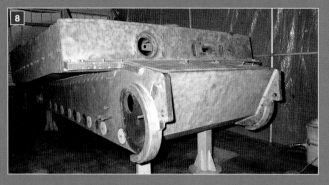

8 The hull stripped and shot-blasted.

9 The torsion bars and road wheels are stacked up.

10 The hull in red oxide primer with some of the torsion bars replaced.

11 The ball mount for the hull machine gun is removed.

12 Removed items from the tank are placed in stillages and stored.

13 The interior of the turret looks tired before restoration.

14 Complete with wheels, suspension and tracks, the hull is collected from the museum by ABRO's Dave Marsh.

15 The turret with the mantlet removed.

16 Ron Luke runs up the Maybach engine on a test stand.

17 The steering unit showing the clutches.

18 A Terex Dozer moves the Tiger hull back to the ABRO workshops after the engine failure.

19 The turret basket from below.

20 The hull before the return of the turret.

21 The turret and basket are manoeuvred in the ABRO workshop.

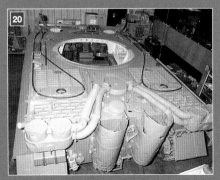

22 When possible, the turret was displayed at the museum.

23 The tank, hull and turret are reunited at ABRO.

24 But this was not the end of the saga. Here, another turret lift is carried out beside the museum.

25 The engine is removed again for further inspection in the Tank Museum workshops.

26 The engine is placed on a test stand in the workshops. On the right are silencers from a Centurion tank that are being used to lessen the noise.

27 Now at the SdKfz Foundation, the engine undergoes another rebuild.

28 Nick Rutherford refits the crankshaft to the new block.

29 The rebuilt engine sits on an adapted pallet ready for testing.

30 The museum's Chieftain recovery vehicle is used to remove the engine after 11 hours of successful running.

31 Ian Aldridge records work carried out – record keeping is essential on such a restoration project.

32 Having disassembled the engine to record any wear, the crankshaft shows minute signs of particles picked up from the white metal bearings.

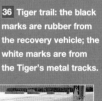

33 Scoring of a cylinder liner can be seen here. This is caused by the liners not seating correctly in the block.

34 Here are the four fuel tanks of the vehicle before they are refitted in the hull.

35 The engine cooling fans fit either side of the main engine compartment. The radiators are installed behind the fan units.

36 Tiger trail: the black marks are rubber from the recovery vehicle; the white marks are from the Tiger's metal tracks.

Restoration: the workshop view – Mike Hayton

After a career in electrical engineering, Mike co-owned a transport business and restored rare French classic cars. He volunteered at the Tank Museum in the early 1980s and came to work fulltime in 1997. He was appointed Workshop Manager in 2005.

The Tank Museum workshop was given responsibility for the Tiger after its return from restoration in 2004, although it was May 2005 before we felt the tank was truly roadworthy.

As has been mentioned, the original engine had been sectioned, so the only alternative was to use the later HL230, of which several were on display in the museum.

Before any work could be undertaken the operating manual was translated from German to English. This gave us a good insight in what was to be a somewhat complicated trip. Many hours were spent setting the tank up 'to the book'. For example; automatic gear changing, timing of clutch operation, 'blipping' the throttle, clutch retardation, and actual gear changing are critical. A small group of volunteers took on the role of setting the various control rods etc to the correct positions, all of which required very accurate adjustments.

During research, it also became apparent that compared to other tanks the museum looks after, the crews and engineers of Tiger 131 would require supplementary training with regard to maintenance and extra skills.

Engine troubles

From the beginning, the engine was the Achilles' heel of the tank. Two engine blocks had been tried without success before Tiger 131 returned to the museum from its restoration. The first suffered catastrophic failure and the second had serious frost damage. A third engine block was sourced, but that too had frost damage. It was sent away for heat treatment, whereby the block was heated to cherry red, repaired and then cooled. The engine was rebuilt and fitted to the tank.

During the launch, coolant leaked from just about every orifice, engine oil pressure was unacceptably low, plus discoloured paint on the left-hand exhaust silencer, together with a misfire, showed signs of the left bank of cylinders running hot, perhaps because of retarded engine timing.

The engine was subsequently taken out by Tank Museum workshop staff and part stripped for inspection. The results were not good.

- Number 1 shell bearing had all but disappeared.
- The crankshaft was beyond repair.
- Even worse, after careful measuring, the block appeared to be warped.
- Coolant had leaked from many areas between both cylinder heads and block, appearing to confirm our findings.
- Metal particles in the engine oil and oil galleries.

We took the view that unless the engine block was in perfect condition it wasn't worth the risk of potentially damaging other mechanical components. Possibly the engine would have to be reassembled, and placed in the tank as a static exhibit.

The SdKfz Military Vehicle Foundation heard of our plight and offered an unused HL230 block – plus the offer of help to rebuild the engine to working order. I took the engine to the SdKfz premises and assisted in taking it to pieces to the last nut and bolt.

After reading operational reports it was clear that small coolant leaks were common on these engines. For example, the pistons ran in liners which could easily be replaced instead of having to re-bore the engine. The increase in cubic capacity from 21 to 23 litres meant that the liners were located extremely close together.

BELOW Brake assemblies removed from the tank.

At their base neoprene 'O' rings sealed the liners to the block with copper rings and gaskets between the engine block and heads. There was a consensus of opinion that these areas were particularly weak, accentuated by the fact that the engine was physically short in order to fit in the Tiger's engine bay.

New improved pistons and liners were brought in from America and fitted to the replacement block together with the original mechanical components. Wartime manuals were available to assist in the rebuilding process, especially with regard to a collar fitted to the crankshaft. If the wrong one were fitted, number 1 big-end bearing would be starved of oil. A correct one was sourced from an engine, retrieved from a river bed.

I was privileged to assist in the last phase of the rebuild, and to see the mighty Maybach roar into life while perched on a pallet. The engine was fitted in time for Tankfest 2006 and it performed faultlessly during the two-day event.

By 2010 the engine had run a total of 11 hours, including the obligatory 15-minute warm-up period each time the tank was driven. Even though museum vehicles don't travel very far, it is workshop policy to carry out a systematic overhaul programme.

During restoration certain modifications and omissions had to be made which compromised the vehicles originality. So as part of the overhaul, the tank would be put back to as near original specification as possible. As a precaution the engine was removed and disassembled to check for wear and tear.

ABOVE LEFT The transmission once removed.

ABOVE The transmission has been exposed in this photograph of a demolished Tiger. This tank was from the 501st Heavy Tank Battalion and was in fact the first Tiger to fall into Allied hands in North Africa. After initial analysis in situ it was destroyed to prevent removal, both sides claiming the demolition in their reports on the incident.

FAR LEFT The rebuilt engine is fired up outside the vehicle. (*Mike Hayton*)

LEFT Dropping the engine back inside the hull takes patience and accuracy. The engine bay is incredibly tight.

Generally the news was good, although:

- A number of rubber 'O' rings between the inlet manifolds and heads had broken up.
- Gas leakage through the head gaskets, caused a weak petrol/air mixture on some cylinders.
- The crankshaft was slightly out of line, resulting in slight wear on some of the journals.
- Excess wear on the electric starter pinion.
- Some soldered joints on the carburettor floats needed attention.

Therefore we feel that having run the engine for 11 hours the wear discovered justified our decision to return the vehicle to running order.

As part of current work we also looked at the following areas:

Electric starter

The electric starter has proved troublesome by not disengaging quickly enough once the engine has started. The starter is hidden from view, plus the interior mechanical noise in the vehicle is so horrendous that it is impossible to tell whether or not the starter has disengaged. To address this problem in order to ensure the starter has disengaged, one of the bolts securing the starter inspection cover is temporarily removed from the bulkhead, so that a visual inspection can be carried out while the starter motor operates.

RIGHT The petrol tanks ready for refitting in the side panniers.

Inertia starter

At the museum we would need at least four people to start the tank using the following method. Driver, petrol primer person/fireman and two people to turn the starting handle.

Because there are many gears inside the inertia starter the operation gets off to a slow start, and requires immense effort. When enough speed has been obtained the small hand device is pushed towards the rear hull plate to engage the starter. This action is repeated several times in order to raise oil pressure. With oil pressure at an acceptable level, the driver turns on the ignition, and the process is repeated. The engine should start immediately.

Tiger 131 was not fitted with the inertia starter during the earlier restoration. In 2010 a complete inertia starter system was constructed using the original drawings, and is now fitted to the vehicle.

Wiring

During the restoration process the tank was rewired using modern screened cable. The museum still has the original wiring looms, which are in remarkably good condition still in their protective steel tubing.

Fuel tanks

When Tiger 131 was first restored, only one of the four fuel tanks was connected and the right-hand reserve tank left out. Some 50 per cent of the pipe work plus the two fuel valves were missing altogether. Re-manufacture of these components was the only alternative in order to revert to original specification, and this pipe work is now fitted.

Fan drives

During restoration of Tiger 131 temporary cooling fan drive units were constructed (see *The engine and auxiliaries*, page 96), these became unreliable and leaked oil. Fortunately an original drawing of the HL230 fan drive gearboxes exists and various companies were approached in order to see if re-manufacture to the original specification was a viable option. Remaking such complex items is no easy task and we have been fortunate in gaining further grants from the Heritage Lottery Fund and the PRISM Fund to help us complete this work.

The vexed question of paint – David Willey

The paint scheme on the Tiger is a small but obviously very visible feature of this historic machine. The story behind the painting of the tank is an interesting example of how originality in a mechanical item can become such a confusing issue.

Paint schemes on German vehicles have for years been the modeller's delight and bugbear. Only in relatively recent years has the serious study of paint schemes as issued in official German orders been studied, mainly through the pioneering work of Tom Jentz and Hilary Doyle. Their research into surviving German production records and orders has led to much more specific knowledge of the dates and actual colour numbers used.

Paint schemes

The Tiger I at the Tank Museum (officially identified as Pz.Kpfw.VI H Ausf. H1, *Fgst Nr* 250122) was completed at the Henschel assembly plant in Kassel on or about 10 February 1943. The chassis serial number plate was not stamped with the letters 'tp', indicating that this Tiger was not prepared for tropical deployment at the assembly plant. Other vehicles in the museum's collection such as the Panzer III have such stamping. During early phases of the restoration project, a documented examination of the layers of paint was not made, therefore systematic proof of the original colour of paint on this Tiger when it left the assembly plant is not available. However, useful evidence of paint was found later in the restoration project.

This Tiger was issued to 504th Heavy Tank Battalion in February 1943. This unit, formed under *Major* August Seidensticker had only been constituted on 8 February. Originally destined for employment in Russia, the 504th was then ordered to prepare for tropical employment on 17 February 1943. Moved first to a staging area in Sicily, elements of the 504th were shipped to Tunisia in March and April 1943.

The RAL system

Germany's paint identification and numbering system RAL (**R**eichsausschuß für **L**ieferbedingungen und Gütesicherung)

was started in 1927. Different government departments such as the Post Office and Army had numbered three-digit paint code sets for their departmental use. The colours were all made from materials readily available within Germany, and as might be expected, made to a high standard. They were created from natural earth pigments so that they would not oxidise or fade – as they were already at base level. In 1941, a single new four-digit system was introduced (previously the Post Office series might have an identical three-digit number to the Army series but for a different colour). Some colours were completely deleted and others marginally changed in their mix. But colour swatches were created in 1941 on enamel-backed cards and these can still be used to give accurate colour matches.

German camouflage experts advised that vehicles needed to be painted in a darker shade than the average background colour or they would show up lighter if painted the same shade (and therefore be more easily located).

An order dated 17 March 1941, stated that equipment employed in Africa was to be painted in a two-tone scheme with two-thirds of the surface coated in RAL 8000 (*Gelbbraun*) and one-third in RAL 7008 (*Graugrün*). A new general order, a year later, dated 25 March 1942, stated that equipment – including

BELOW Tiger 131 in the Tank Museum, looking worn and (therefore) authentic before the major restoration began. However, the paint scheme is totally spurious.

RIGHT Testing the paint scheme. The two-tone scheme caused much comment when first revealed on the restored vehicle.

RAL 7008 (over)

RAL 8000 (over)

RAL 8000

vehicles – with units employed in Africa were to be painted in RAL 8020 (*Braun*) and RAL 7027 (*Grau*). These new colours were to be requisitioned through normal supply channels; however the previous colours RAL 8000 and RAL 7008 were to be used up first.

Therefore, it might be expected that a Tiger employed in Tunisia in early 1943 should have been painted RAL 8020 and RAL 7027. But, no physical evidence has been found to support this case. The other remaining Tiger captured by the British in Tunisia (No 712) that was subsequently sent to the US Army's Aberdeen Proving Ground in Maryland, USA, for analysis also shows evidence of the earlier paint scheme.

During disassembly of the Tiger for restoration, two colours were found on surfaces that had been protected from subsequent applications of paint. The colour RAL 7008 was found between road wheels and on the lower hull side and the colour RAL 8000 was found under the stowage bin on the rear of the turret.

Prior to being captured in Tunisia, this Tiger had been painted in a two-tone scheme of RAL 8000 (*Gelbbraun*) and RAL 7008 (*Graugrün*). On close inspection of the better-quality photographic prints taken of the tank soon after capture in Tunisia, the two-tone paint scheme can be easily discerned on the turret sides, turret stowage bin, driver's front plate, and right upper hull side. Due to dust, light effects, oil marks and the similarity of the grey tones from RAL 8000 and RAL 7008, it is not possible to determine the pattern on all surfaces but there is enough evidence to provide a sample and style to the patterning. The vehicle was returned to this scheme as part of the restoration process.

The paint schemes added in the museum after the tank's gifting in 1951, including the overall 'sand' colour and large identification numbers, has become part of the general misinformation concerning German vehicles that the Tank Museum has inadvertently perpetrated over the years. It also led to many questions when the subsequent correct scheme was restored to the vehicle.

What we see and remember, especially from

RIGHT Another Tiger 131 – this vehicle was part of the 501st Battalion in Tunisia. The turret numbering system indicated the company, the platoon and the tank number in the platoon. Each Tiger battalion had a different numbering style that now helps to identify the individual unit.

many years ago, often becomes a given truth – when in fact, what we may have been seeing was inaccurate or a myth. As clear evidence of this we asked many enthusiasts what the numbering on the Tiger should look like. Large red numerals, with a lighter outline, are what most suggested. However, this scheme was the later addition that many saw on the Tiger in the museum. Returning to the evidence supplied by the early photographs taken soon after capture, the numbering is in RAL 3000 (flat red as per orders) with the numerals of modest size and not outlined in any other colour.

Photographic evidence also shows that the Tiger had been repainted in parts prior to being examined by the King and Prime Minister Churchill in Tunisia. The shield of 1st Army and the diabolo of 25th Army Tank Brigade had been added to the front plate as marks of conquest. After being shipped to England, it was repainted during its examination at the School of Tank Design and again when displayed at the Tank Museum.

It is important before taking information from images to know the date of the pictures being studied. In the Tiger's case, information should only be gleaned from the initial photographs taken directly after capture. It is also important not to judge actual colours from black and white images unless there is clear supporting evidence. The relative tonal values of a black and white image when compared to one in full colour can appear very different. Recent evidence has shown how a well-known and researched vehicle such as Rommel's *Kfz* 21 Horch 901/40 *Kommandeurkabriolett* which, in contemporary black and white images, looks as if it was painted in some form of desert paint scheme, actually spent its service life in the standard *Wehrmacht* grey (RAL 7021).

The varying camouflage schemes that can be detected on German armoured vehicles have an underlying palate of colours that are dependent on the current orders issued. Most vehicles were given an initial paint scheme at the factory. This base coat was then over-painted with approved colours to a unit's individual scheme on arrival. Paint was delivered in concentrated paste form that could be mixed with water or petrol for application.

Images show crews painting their vehicles in the field using air compressors and spray cans. Inevitably, captured paint was used in circumstances where regular supplies were difficult to come by (such as in North Africa), and where weather and local conditions led to the improvisation of schemes with whitewash or mud. However, the general assumption that any paint to hand would be used has to be viewed with caution. Not only did the German system and mentality frown on any departure from the set procedure and orders, but Europe, Russia and North Africa simply did not have the paint available in the quantities and from the outlets that we are used to today. The world was a much less object-rich environment in those days and the hardware supermarkets we are familiar with simply did not exist.

LEFT The Kfz21 Horch 901/40 *Kommandeurkabriolett* that was used by Rommel in North Africa. Its story of recovery, identification and restoration is worthy of its own book, but for this volume it provides an excellent example of the dangers of trying to judge colours from black and white images. The vehicle spent all its service in RAL 7021, the *Wehrmacht* grey, yet looking at images of the vehicle in North Africa the vehicle looks as if it were painted in a much lighter colour.

LEFT The restored vehicle has been finished in the original paint scheme, RAL 7021.

Restoration: the volunteer's view – Stevan Vase

Stevan Vase has a background in general engineering but he ended up in IT as a project manager, working for one of the top four IT service providers in the UK. He has been a Friend of the Tank Museum and a volunteer in the workshops for around 15 years.

I started off doing the basics, as every volunteer does, the nonetheless important tasks of fetch and carry, POLs (Petrol Oil Lubricants) during Tankfest and workshop weekends. Under the watchful eye and tutelage of Mike Hayton I found myself being moved on to ever more complex mechanical tasks, regularly doing Chieftain gearbox and engine changes, which is highly enjoyable with the rain coming at you sideways. With trust that had been earned I was asked to assist the workshop staff in restoration and ongoing maintenance on some of the museum's rarer exhibits, such as assisting on the repair and rebuild of the Panzer III engine bay.

Without my knowledge all along I was being assessed by Mike, Chatty and the rest of the workshop crew. This came to a head in 2004 when Mike asked me to accompany him to his office; I was at that time assisting with the conservation of a Mk III Chieftain, Tiger had just come back to the museum from ABRO and I had been casting an interested eye over it. While walking to the office I wondered how much

trouble I was in. Mike thanked me for my hard work and asked me a couple of questions about my family; I thought I was out on my ear for good. He asked me if I would not mind stopping work on the Mk III Chieftain to become part of the Tank Museum team that look after Tiger, keep it running and continue the restoration and conservation of the vehicle. I was stunned but promptly bit his arm off with a YES!

The research for me started by reading every Tiger book I could lay my hands on and obtaining copies of the vehicle's manuals. One of the first problems we had to resolve was the power turret traverse. Needless to say this is an overly complex system that is driven from the gearbox through a cone clutch, which is engaged via a handle in the radio operator's compartment, into the traverse gearbox in the bottom of the hull and then up into a hydraulic two-speed motor which then powers the traverse driveshaft. The power traverse worked intermittently and made some fairly nasty grinding noises. Fearing the worst we worked our way up the traverse drive chain checking for play, emptying the various gearboxes in the traverse power and manual drive trains looking for metallic shiny lumps in the oil but found nothing. It came down to the distance between the turret basket floor and its height off from the top of the traverse gearbox in the centre of the hull. There is a male shaft that fits in a female coupling on the traverse gearbox in the hull and the hydraulic motor mounted on the turret basket floor. This shaft was not engaging properly. A few adjustments to the turret basket were made and the shaft seated properly in the female couplings. We successfully tested the power traverse and timed it on fast speed setting with the engine revs at 1,500rpm; it took 60 seconds and managed to resolve ongoing discussion on various forums overnight on how fast a Tiger turret could turn.

During the latest stage of the ongoing restoration we have been looking at all of the vehicle's subsystems and ensuring that they are restored to the correct period configuration for a Tiger I that has had an HL230 upgrade. We have found that each of the subsystems that make up the vehicle are very complex for the

BELOW Stevan Vase is one of the Friends of the Tank Museum who volunteers in the workshops. He is seen preparing the sectioned Maybach HL210 engine for display.

FAR LEFT With the turret removed, stowage for some of the 88mm rounds is revealed.

LEFT Replacing the turret on Tiger 131 outside the Tank Museum.

period, and engine and gearbox temperature are critical to the operation of these systems and their reliability. While researching the fuel system we have found that it has an electronic auxiliary fuel pump that is mounted in the engine compartment; there is no mention of this in the various operators' and workshop manuals we have. It is, however, mentioned in the School of Tank Technology reports.

There are many more ongoing restoration projects that will keep both me and the museum busy for many years to come. From working closely on the vehicle, I think I need to dispel a few myths about the Tiger: it is not the slow lumbering beast portrayed in many books and films; it has the same mobility and ground pressure as a Panzer IV; the vehicle is very agile and when driven correctly it is quite fast for its size and age.

The vehicle is overly complex, although operation of the subsystems correctly will ensure the major components operate to the correct service life. The main issue with the vehicle is that, under pressure and combat, these complex operating procedures were liable to break down easily, with soldiers taking short cuts to get the vehicle up and running again which, in turn, led to high breakdown rates. Also, the Germans were trying to push the envelope of materials technology for the period. Some of the materials they used were on their limit for the mechanical systems that make up the vehicle, final drives being a particular weak point.

I have had the opportunity of crewing the vehicle in many of its displays at the museum; one of the highpoints was manning the gun in a re-enactment battle at Tankfest; I was operating the power traverse while looking through the TZF 9b (*Turmzielfernrohr* 9b) gun

sight tracking an M18 and a Sherman while firing pyros at them. It was like a real-life version of a tank simulation.

I do have to pinch myself sometimes as I am very honoured and privileged to be given the opportunity to undertake work on this precious historic exhibit. It's an even greater honour to meet the veterans who were there when it was captured and hear their stories. We can count ourselves lucky that the Nazis designed themselves out of the war-winning numbers production race, by ensuring that systems they designed were as near mechanically perfect as they could be, rather than designed for mass production.

My hope is that we can keep this vehicle running as a fitting tribute to all those who served in the Second World War and ensure that future generations can interpret the vehicle through seeing, feeling, hearing and smelling the distinct sensual assault of a Tiger rumbling past them at the museum for many years to come.

There is a dichotomy of 'to run or not to run' and the museum is undertaking a lot of advanced research in this area. It is positioning itself as a centre of excellence for the conservation and preservation of AFVs, with a number of postgraduate research students from leading British universities looking into issues such as storing non-running vehicles and tribology (the science of friction and wear). It has formed a group of leading academics and representatives from other major AFV collections to help set standards of care and ethics for AFV preservation. This research will better inform the way we look after the Tiger and all of the vehicles in the collection, thus consolidating the position of the Tank Museum as having the best collection of tanks in the world.

'When changing gear in a Soviet T-34, a second crew member would often need to assist the driver on the gear stick. Changing gear in the Tiger can be done with a single finger. In addition to that, since the Tiger uses a steering wheel, it's very easy to change gear and steer at the same time.'

Darren Hayton, Tank Museum volunteer driver

Chapter Five

Running the Tiger

Mike Hayton

Keeping the Tank Museum's tanks in working condition is a challenge. Every vehicle is different – some are easy to work on, while others test even the most skilled engineer. In rising to these challenges the museum's dedicated workshop staff need to be detectives, multilingual and, above all, patient. In this chapter you can gain an insight into what it is like to operate the Tiger I.

OPPOSITE The Tiger in the workshops dwarfs the Swedish Stridsvagn M40/L.

Health and Safety – the golden rules

- Carry out a risk assessment for the task.
- Always use step ladders to get on and off tanks (the Tiger is 2.8m high).
- Always wear PPE (Personal Protection Equipment), for example safety glasses, gloves, safety boots etc.
- Cordon off the area surrounding the tank, if required.
- Have the correct fire extinguisher at hand.

Keeping records

The workshop has a system for recording vehicle use and work done. This applies to all vehicles in the museum – from main battle tanks to the fork lift truck. Each vehicle has its own folder containing:

- Mobility sheet – First Parade, Last Parade/distance travelled etc.
- Work done – By whom/time taken/materials used.
- Lubricants – Record of the type of lubricants/capacity/amount used.
- Paint details – Colour details/when last painted/markings.
- Photographs – Details of work done /colour changes/annual vehicle record.

Before starting up (First Parade)

Make a visual check of the vehicle – tracks, wheels, idlers, track tensioners, sprockets, track link pins, oil/coolant leaks, and suspension.
- Open hatches and check for security; check internally for fuel odour and hydraulic leaks.

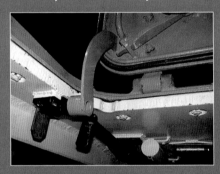

- Check gearbox oil level using dipstick.
- Check all controls for operation.
- Remove the *Feifel* air ducting from above the engine deck.
- Turn the turret manually so the gun barrel is at right angles to the hull.

- Request the assistance of four burly men to help raise the rear engine deck (which is hinged and held open by a hook).
- Carry out a general check inside the engine bay.
- Check the oil level in the air filter bowls.
- Check oil levels in the fan drive gearboxes.
- Check the engine oil level using the dipstick; this is a cursory check to establish there is enough oil in the oil tank to run the engine.
- The oil level should be at or above the 'low' mark.

- Check coolant levels by undoing the screw caps on each of the radiators.
- It is now safe to go through the start procedure.

Start-up
- With the engine running at 500rpm check:
- Engine oil level, top up if necessary.
- Gearbox oil level, top up if necessary.

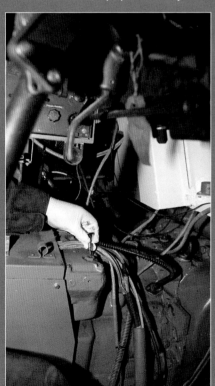

Run the engine at 2,000rpm check:
- Engine oil pressure should be at least 3.5 atu (3.5 bar).

Run the engine at 1,000rpm check:
- The battery charge warning light should be out.

After the warm-up period the deck is dropped into position (by four fit men), the turret set to straight ahead and the lock applied. Replace the Feifel air ducting. The tank can now be driven from the museum hall to the workshop, where:
- The gearbox oil will be topped up if necessary.
- The engine oil will be changed, and the filter cleaned.
- Further checks made regarding coolant and fuel leakage.

At this point the relevant paperwork can be bought up to date.

After activity (Last Parade)
Almost a reversal of the First Parade, but with the following additions:
- Top up the oils while the engine and gearbox are hot.
- Clean the tank thoroughly inside and out.
- Complete records.

Driving Tiger 131 – Darren Hayton

Darren Hayton works in The Netherlands as an astrophysicist specialising in astronomical space instrumentation. He has been a volunteer at the Tank Museum for over 17 years and has closely followed the restoration of Tiger 131 from the beginning. He is particularly interested in the engineering design and the operation of this vehicle and captioned the original German wartime manual drawings that appear in this book.

First impressions

From the moment you get in through the driver's hatch you realise that the Tiger is unlike any other vehicle of its era. Firstly, the hatch is offset from the seat making it near impossible to drive the tank with your head out of the vehicle. Next is the layout of the controls. Uniquely, the Tiger controls closely resemble a modern-day car or truck in that it has a steering wheel, accelerator/brake/clutch pedals, a handbrake and a gear selector in the places that you would expect for a left-hand-drive vehicle. This makes driving the Tiger very intuitive for a beginner. It's also extremely helpful for driving in a closed-down configuration (necessary for a live turret). With the hatches closed it is essentially dark inside the vehicle so you have to be able to reach for controls without seeing them.

First impressions when sitting in the driver's seat are the lack of visibility and, when running, the overwhelming mechanical noise produced by a long chain of driveshafts and straight cut gears. Next comes a feeling of great responsibility being in charge of such a significant and historic machine. Tigers have a reputation of poor reliability so our goal with Tiger 131 is to operate it in the most mechanically sympathetic way possible. That has involved many hours reading books, reports and manuals in order to gain an understanding of how a Tiger works and therefore how best to drive and maintain it to achieve the lowest wear and tear. Many other idiosyncrasies of this vehicle have been figured out and recorded along the way. For me, what stands out is the effort that the designers went to (mainly with transmission and steering) in order to reduce physical effort required by the driver – even to the extent that the resulting transmission was probably a step beyond the technology available at the time, which must have contributed to the tank's reliability issues. An example of this is when changing gear in a Soviet T-34, a second crew member would often need to assist the driver on the gear stick. Changing gear in the Tiger, however, can be done with a single finger. In addition to that, since the Tiger uses a steering wheel, it's very easy to change gear and steer at the same time.

Operation

Running Tiger 131 is not a simple exercise. The tank is stored in the museum with no batteries and minimal petrol. Several hours before running, batteries are installed and fuel is added. An estimate of 2gal/mile is used when determining the amount of fuel needed. A detailed walk around the vehicle follows with particular attention paid to tracks, wheels, bodywork and stowed equipment. Levels are then checked for coolant, transmission/steering oil, final drive oil and hydraulic oil. Engine starting is a minimum four-person job needing a driver, commander, fireman and someone to guide. The commander is responsible for fuel control (stop valves and priming), electrical master switch and operation of the oil pressure bypass valve during start-up. After start-up they assist with the driver's visibility and guidance. An intercom links all members of the Tiger crew. Being a petrol engine, fire is always a risk that is taken very seriously. The driver and commander each have a fire extinguisher in addition to a dedicated fireman outside of the vehicle. Finally, a guide person stands directly in front of the tank at all times for signalling direction during tighter manoeuvres and reversing.

BELOW Workshop staff member Ian Aldridge prepares the Tiger for an inertia start. Ian has played a pivotal role in the continuing restoration and conservation of this vehicle.

Engine start-up procedure

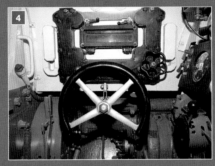

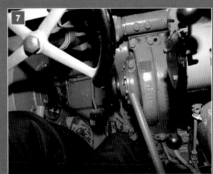

1 Electrical master switch set to ON.

2 Fuel stop set to ON.

3 Ignition key inserted and turned. Oil pressure bypass set to allow carburettors to fill, fuel primer used to pump neat fuel into the intake manifolds. Fire extinguishers ready.

4 Check steering wheel is straight.

5 Check transfer box is in neutral.

6 Check handbrake is on.

7 Check clutch pedal is depressed.

8 Three pumps of the primer (from cold start).

9 Start button pressed.

If the procedure has been carried out correctly then the engine starts instantly and idles at 500–600rpm.

From now on, while the engine is running, regular checks of the oil pressure gauge are essential. At the same time the final level (engine oil) can be checked as it is a dry sump engine. *Tigerfibel* states that for a cold start the engine should be idled until the transmission has warmed up enough that it feels warm to touch (10–20mins). We have found that this is essential for a smooth gear change.

Driving the Tiger

1 Driving is relatively easy. The clutch is only required for starting and stopping, but care must be taken not to rest a foot on the clutch pedal during driving as the clutch is hydraulically assisted (another Tiger first) and pressing it even a few millimeters will cause it to slip or even fully disengage drive.

4 Steering is two-stage in that there are two turn radii available – half and full lock. The turn radius is also dependent on the engaged gear, whereby a tighter turn is given by a lower gear. By design, it is not possible to fully lock a track with this tank.

2 The gear lever can select (1–8 forward, or 1–4 reverse). Changing gear is achieved by moving the gear selector to the desired gear and then pushing the selector to the side to implement the change. This then automatically disengages the clutch, revs the engine a little, engages the gear and then re-engages the clutch. This all happens using hydraulics and without the need for any further input from the driver.

3 Engine rpm has been shown to be critical for smooth gear changes. If the engine speed is too low or too high then the resulting change will be very hard and the vehicle will lurch – even at 57 tons!

5 Neutral turns (i.e. one track driving forward and the other driving in reverse) are possible with the Tiger when the transfer box is in neutral. A black button is located on the side of the transmission, which locks the transmission ensuring that a true neutral turn is achieved.

Driving impressions

The Tiger isn't as challenging or physically demanding to drive as many of the other vehicles of its era. However, it still takes practice to be able to drive it smoothly and it certainly lets you and the rest of the crew know when you don't get it right.

I've been extremely fortunate to have had the opportunity to learn to drive a Tiger I over the past four years. We continue to learn new things about this tank and improve upon our procedures each time we run it. With due care and consideration I hope that Tiger 131 will be operational for many years so that future generations may observe and experience this historic tank.

My most memorable driving experience in the Tiger was during simulated battle as part of a Tank Museum special event. The Tiger turret was used for the first time, meaning that the vehicle was operated with all of the hatches closed. Sitting in the dark with a full crew, the noise and fumes from the engine, and with large pyrotechnic charges detonating just a few metres away put us all a step closer to appreciating what actual Tiger tank crews had to endure for real. You can only imagine how much worse it must have been for Allied crews who had to face the Tiger.

Wheel and track maintenance

Maintenance of the tracks is of paramount importance. It is vital that wear and tear is kept to a minimum by gentle use and correct track tension.

Wheels

Tiger I has a total of 48 road wheels, each one fitted with 800/94D Continental tyres. During the track inspection process all 48 wheels are checked, including wheel nuts. Special attention is paid to high wear rate areas such as where the inner wheels come into contact with the track horns.

Track inspection and tensioning

At the museum three people will be required in order to carry out this task: driver, guiding person and radio operator. At the same time a minute inspection of each track link takes place. To adjust the right-hand track for example:

1 Carry out engine start procedure.
2 Place the tank on hard level ground.
3 Confirm the direction lever is in neutral.
4 The person guiding will indicate for the Tiger to prepare to move, the radio operator will tell the driver to carry out a neutral turn by turning the steering wheel to the right, apply the hand brake and immediately return the steering wheel to the central position.
5 By carrying out this manoeuvre, the left track is under tension, while the upper right track sags above the road wheels.
6 If the track is adjusted correctly it should be just touching the second road wheel.
7 Adjustment is carried out by removing armoured covers at the rear of the tank. A special tool is inserted through the aperture and turned clockwise to tension the track, or anti-clockwise for de-tension.

LEFT Correct track adjustment.

LEFT Location of track adjusting mechanism.

'Although there is a general grouse that the V-12
HL230, 21-litre Maybach engine is underpowered for
the Tiger I tank, there seems to be no real evidence for
it because there are few major engine breakdowns and
the AFV is claimed to have a good turn of speed in all
gears. The root cause would appear to be short engine
life owing to overloading when used for towing, but
while it lasts the engine gives all that is asked of it.'

Royal Armoured Corps liaison letter
August 1944

Chapter Six

The Maybach Engine

Mike Hayton

A 57-ton tank will go nowhere without an engine. Much has been written on the perceived weaknesses of the Tiger's engine, the famous 12-cylinder 700hp Maybach. Post-war analysis attributed many engine malfunctions in German wartime tanks to the poor quality of wartime oil, but good drivers and mechanics could keep a Tiger running with reliability similar to other German tanks like the Panzer IV and the Panther.

OPPOSITE The Maybach engine is slotted back into its narrow bay.

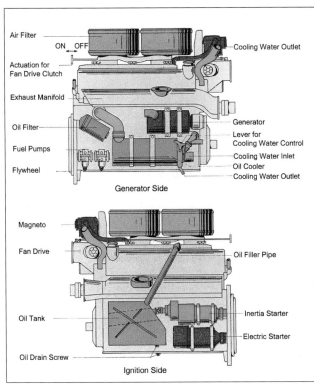

Air Filter

ON   OFF

Actuation for
Fan Drive Clutch

Exhaust Manifold

Oil Filter

Fuel Pumps

Flywheel

Cooling Water Outlet

Generator

Lever for
Cooling Water Control

Cooling Water Inlet

Oil Cooler

Cooling Water Outlet

Generator Side

Magneto

Fan Drive

Oil Tank

Oil Drain Screw

Oil Filler Pipe

Inertia Starter

Electric Starter

Ignition Side

The engine and auxiliaries

The original engine fitted to Tiger 131 was a Maybach HL210 TRM P45, which was sectionalised many years ago. It had an aluminium engine block as opposed to cast iron, fitted to the later HL230. The first 250 vehicles to be produced were fitted with the HL210, which was deemed to be underpowered. It became standard practice to substitute the HL230 if an engine change was to be undertaken. In order to fit the more rigid HL230, minor alterations within the engine bay were required including changing the way the engine was connected to the radiator cooling fans.

The HL210 fan drive consisted of a shaft driven from the rear of the engine which powered a gearbox fitted centrally at the rearmost end of the engine compartment.

Two driveshafts then travelled at right angles from the gearbox in order to drive two

LEFT Maybach HL230 engine.

BELOW Front and rear views of the engine.

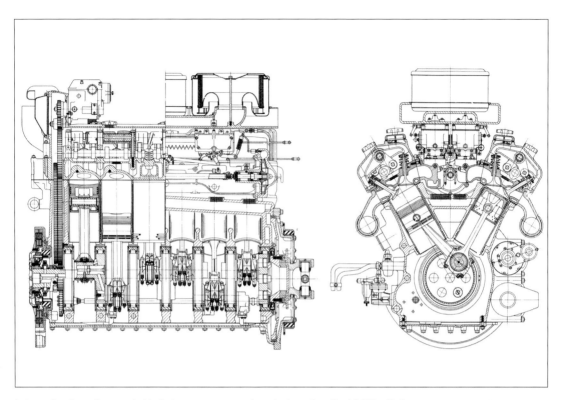

twin cooling fan units mounted in their own compartments to the right- and left-hand side of the engine bay.

Basic engine specifications

HL210 TRM P45	HL230 TRM P30
V-12 water-cooled	V-12 water-cooled
aluminium block	cast-iron block
Capacity	
21.33 litres	23.88 litres
Power output	
650bhp @ 3,000rpm	700bhp @ 3,000rpm

Terminology: HL – High performance; TRM – Dry sump with impulse magneto; P – Specifically designed as a tank engine.

Note: even though the HL230 produces 50bhp more than the HL210, engine change instructions say that no increase in carburettor main jet sizes are required.

In order to replace the HL210 with the HL230 an in-field conversion kit/procedure was supplied to the engineers, including changes to the way the fans were driven.

The HL230, however, has two output shafts, located one on each side of the engine. These connect via two small gearboxes mounted directly to the cooling fan housings. Inside the gearboxes are bevel gears plus two oil pumps, one to lubricate the internal gears, the other to force oil along channels in order to lubricate the cooling fan bearings.

Engine fixtures and fittings

An interesting feature of the HL230 is a valve unit fitted to the upper front of the engine. When the crankshaft revolves, petrol from two main engine fuel pumps is piped via the valve to four Solex twin-choke carburettors sitting in the centre of the 'V'. Engine oil is also pumped through a separate chamber, which, when the oil pressure to various engine mechanical components has built up sufficiently, opens the valve to allow petrol through to the carburettors.

ABOVE Cutaway views of the Maybach HL230.

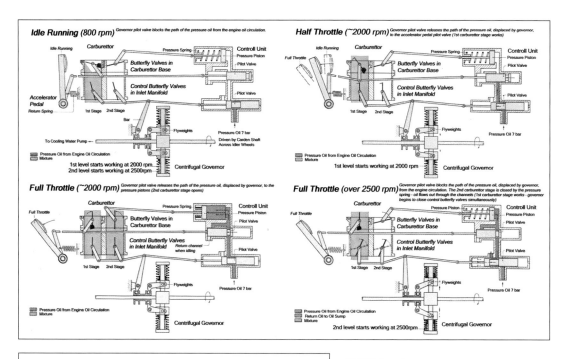

Idle Running (800 rpm) Governor pilot valve blocks the path of the pressure oil from the engine oil circulation.

Idle Running
Carburettor
Pressure Spring
Controll Unit
Pressure Piston
Butterfly Valves in Carburettor Base
Pilot Valve
Control Butterfly Valves in Inlet Manifold
Pilot Valve
Accelerator Pedal
Return Spring
1st Stage 2nd Stage
Bar
Flyweights
Pressure Oil 7 bar
To Cooling Water Pump
Driven by Carden Shaft Across Idler Wheels
Pressure Oil from Engine Oil Circulation
Mixture
1st level starts working at 2000 rpm
2nd level starts working at 2500rpm
Centrifugal Governor

Half Throttle (~2000 rpm) Governor pilot valve releases the path of the pressure oil, displaced by governor, to the accelerator pedal pilot valve (1st carburettor stage works)

Idle Running
Carburettor
Pressure Spring
Controll Unit
Pressure Piston
Full Throttle
Butterfly Valves in Carburettor Base
Pilot Valve
Control Butterfly Valves in Inlet Manifold
Pilot Valve
1st Stage 2nd Stage
Flyweights
Pressure Oil 7 bar
Pressure Oil from Engine Oil Circulation
Mixture
1st level starts working at 2000 rpm
Centrifugal Governor

Full Throttle (~2000 rpm) Governor pilot valve releases the path of the pressure oil, displaced by governor, to the pressure pistons (2nd carburettor stage opens)

Full Throttle
Carburettor
Pressure Spring
Controll Unit
Pressure Piston
Butterfly Valves in Carburettor Base
Pilot Valve
Control Butterfly Valves in Inlet Manifold
Return channel when idling
Pilot Valve
1st Stage 2nd Stage
Flyweights
Pressure Oil 7 bar
Pressure Oil from Engine Oil Circulation
Mixture
Centrifugal Governor

Full Throttle (over 2500 rpm) Governor pilot valve blocks the path of the pressure oil, displaced by governor, from the engine circulation. The 2nd carburettor stage is closed by the pressure spring - oil flows out through the channels (1st carburettor stage works - governor begins to close control butterfly valves simultaneously)

Full Throttle
Carburettor
Pressure Piston
Controll Unit
Pressure Piston
Butterfly Valves in Carburettor Base
Pilot Valve
Control Butterfly Valves in Inlet Manifold
Pilot Valve
1st Stage 2nd Stage
Flyweights
Pressure Oil 7 bar
Pressure Oil from Engine Oil Circulation
Return Oil to Oil Sump
Mixture
2nd level starts working at 2500rpm
Centrifugal Governor

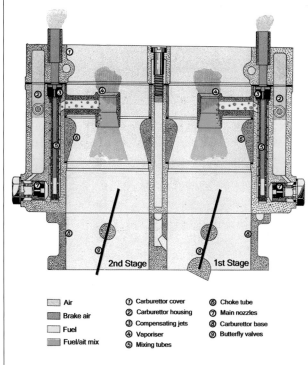

2nd Stage   1st Stage

☐ Air	① Carburettor cover	⑥ Choke tube
☐ Brake air	② Carburettor housing	⑦ Main nozzles
☐ Fuel	③ Compensating jets	⑧ Carburettor base
☐ Fuel/ait mix	④ Vaporiser	⑨ Butterfly valves
	⑤ Mixing tubes	

ABOVE Speed governor.

LEFT Operating mode of main nozzles in twin down draught terrain carburettor 52 J FF II D.

If for any reason the oil pressure drops, the engine is starved of petrol and stops before serious damage is caused.

To the rear of the carburettors are twin magnetos, each one feeding one bank of six cylinders.

On the right-hand side of both the HL210 and HL230 (facing the front of the tank) are an 18.4-litre engine oil tank and oil level dipstick, plus two starter motors – a 24V electrical and an inertia. Both starters are mounted one above the other forward of the oil tank.

Oil tank

Both types of engine have a separate engine oil reservoir. This system is referred to as a 'dry sump'. Instead of storing oil underneath the engine in a sump tray, the oil is contained in a separate tank on the side of the block. This helps to lower the centre of gravity by being able to locate the engine at a lower level in the

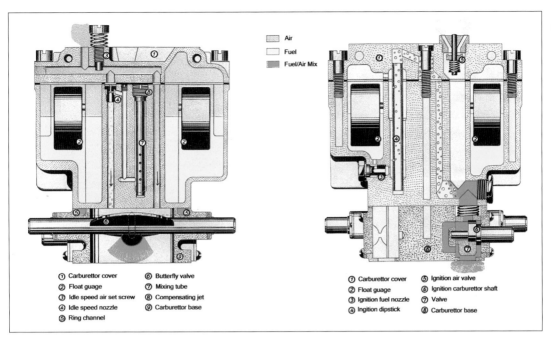

① Carburettor cover	⑥ Butterfly valve		
② Float guage	⑦ Mixing tube		
③ Idle speed air set screw	⑧ Compensating jet		
④ Idle speed nozzle	⑨ Carburettor base		
⑤ Ring channel			

⑦ Carburettor cover	⑤ Ignition air valve
② Float guage	⑥ Ignition carburettor shaft
③ Ignition fuel nozzle	⑦ Valve
④ Ingition dipstick	⑧ Carburettor base

Air
Fuel
Fuel/Air Mix

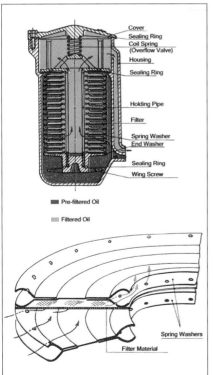

Cover
Sealing Ring
Coil Spring (Overflow Valve)
Housing
Sealing Ring
Holding Pipe
Filter
Spring Washer
End Washer
Sealing Ring
Wing Screw

Pre-filtered Oil
Filtered Oil

Spring Washers
Filter Material

engine bay. It also means whatever angle the tank assumes, the engine oil pressure will remain constant. This is achieved by having three oil pumps. One pump delivers high pressure oil from the reservoir to the engine while two 'scavenge' pumps retrieve oil from under the crankshaft and send it back to the oil tank, where, the whole process starts all over again.

Electric starter

The vehicle batteries are connected in such a way that auxiliaries such as internal lighting etc

ABOVE Idle speed and ignition units in twin down draught terrain carburettor 52 J FF II.

FAR LEFT Oil filter.

BELOW Oil circuit in engine.

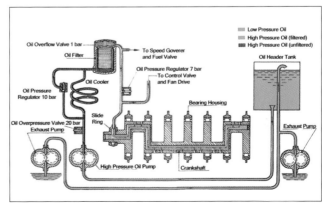

Low Pressure Oil
High Pressure Oil (filtered)
High Pressure Oil (unfiltered)

Oil Overflow Valve 1 bar
Oil Filter
Oil Cooler
Oil Pressure Regulator 10 bar
Oil Overpressure Valve 20 bar
Exhaust Pump
Slide Ring
High Pressure Oil Pump
Crankshaft
To Speed Goverer and Fuel Valve
Oil Pressure Regulator 7 bar
To Control Valve and Fan Drive
Bearing Housing
Oil Header Tank
Exhaust Pump

are connected by a 12-volt supply, whereas the electric starter uses 24 volts, supplied by a change-over relay mounted in the engine compartment.

Inertia starter

In the handbook it states that the tank must be started initially with the inertia starter, using a starting handle, an adapted *Kübelwagen* or a small petrol engine specifically made for the job.

Operation of the inertia starter is quite novel. At the rear of the tank, lower centre, there is a round inspection cover. The cover is first removed and a special plate inserted over the aperture. The starting handle passes through a hole in the plate to a dowel fitted to a sprocket assembly located on the rear engine mounting plate. A chain drive connects to a second sprocket fitted with a keyed shaft. This shaft travels through a tunnel in the oil reservoir tank to the inertia starter via a universal coupling. A series of control rods held together by small ball joints connect the starter to a small spring-loaded device situated adjacent to the starting handle aperture.

ABOVE The rebuilt inertia starter is test fitted to a redundant HL230 block.

BELOW How it would work – the starting handle would fit through the rear wall of the engine compartment and engage gearing.

Generator

The Bosch GUL 1000/12/1000 generator is integrated into the electrical circuit. Cooled by air ducted from the fighting compartment the generator output is controlled by a voltage regulator located in the engine bay.

Heat exchanger

Its primary role is to cool the engine oil by using cooled water from the radiators. Water travels through this unit in its own pipe cooling the engine oil, which is also pumped at high pressure in a jacket surrounding the water pipe.

Electrical

Below the turret floor are two 12-volt 400-amp batteries connected in series/parallel and controlled by a master switch situated on the bulkhead wall. The electrical system is protected by a total of 20 fuses:

- 12 x 15 amp – lighting etc (located on dashboard).

- 2 x 80 amp – 12-volt/24-volt change-over relay.
- 1 x 80 amp – voltage regulator.
- 5 x 15 amp – fire extinguishing system.

Fuel delivery

Petrol is supplied to the engine via two pairs of fuel tanks totalling 534 litres, each pair mounted forward of the radiators and protected by the armoured deck. There are two fuel fillers located just behind and to the right and left of the turret bustle. During replenishing fuel enters via the filler and into the upper fuel tanks, which in turn filters fuel to smaller tanks mounted immediately below. These also act as reserve tanks carrying 70 litres each (enough to travel 30km on a hard surface). Fuel then travels through an elaborate pipe system, via two hand-operated fuel valves mounted on the fighting compartment bulkhead.

Each valve has three positions:
- Position 1 – Main tanks.
- Position 2 – Reserve tanks.
- Position 3 – OFF.

BELOW Fuel system for the HL230 engine.

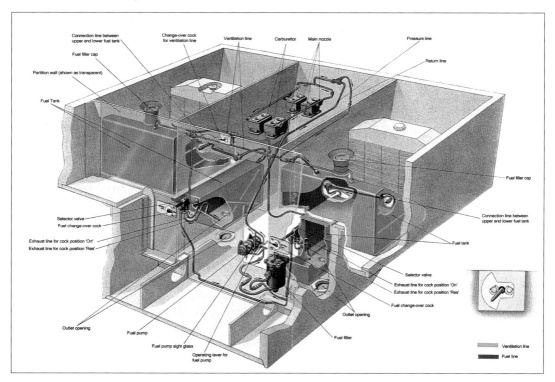

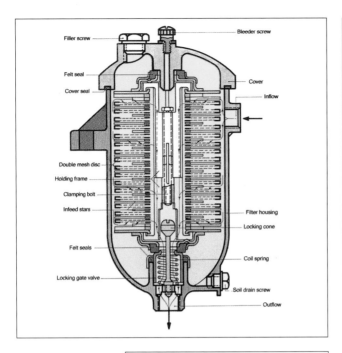

Filler screw
Bleeder screw
Felt seal
Cover
Cover seal
Inflow
Double mesh disc
Holding frame
Clamping bolt
Infeed stars
Filter housing
Locking cone
Felt seals
Coil spring
Locking gate valve
Soil drain screw
Outflow

ABOVE Fuel filter.

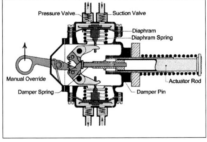

Pressure Valve
Suction Valve
Diaphram
Diaphram Spring
Manual Override
Actuator Rod
Damper Spring
Damper Pin

RIGHT Fuel pump.

BELOW Cooling system.

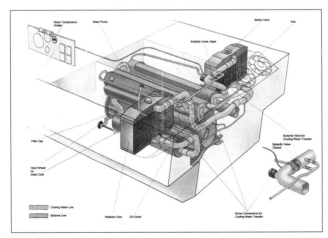

Water Temperature Gauge
Water Pump
Safety Valve
Fan
Butterfly Valve, Open
Filter Cap
Butterfly Valve for Cooling Water Transfer
Butterfly Valve Closed
Hand Wheel for Drain Cock
Cooling Water Line
Balance Line
Radiator Core
Oil Cooler
Screw Connections for Cooling Water Transfer

Fuel consumption

The Tiger ran on petrol and it took 534 litres to fill its four fuel tanks, two on each side of the rear engine bay. That was the equivalent of 27 Jerry cans containing 20 litres each, or 3 x 200-litre drums. One 3-ton supply truck such as the famous Opel Blitz S-Type could carry 110 Jerry cans or 11 barrels of petrol, enough for three Tigers.

Fuel could be poured from Jerry cans, which was exhausting and time-consuming work, siphoned if time was available, or pumped with a small hand-operated or powered pump to transfer fuel from the large drums to the filler points at the rear of the engine deck on each side. Fuel consumption was estimated at 400 litres per 100km travelled on roads, or 700 litres per 100km travelled cross-country. Without further refuelling the range of a Tiger with a full set of fuel tanks was 140km by road or 80km cross-country. This rate of fuel consumption and the wear to the vehicle when driving any distance was one of the reasons rail transport was considered vital to move tanks as near to the action as possible before off-loading.

All four tanks are connected together by a pipe linking the fuel valves. Fitted to the top of each of the main tanks is an air breathing system controlled by a lever situated on the wall of the fighting compartment bulkhead.

From the fuel valves, petrol passes through a filter to the engine fuel pumps, which are driven via the engine timing gear by a shaft which also drives the magnetos. An electrically operated priming pump is located on the engine compartment wall opposite the mechanical pumps (only fitted to vehicles with HL210 engine).

Cooling system

The cooling system holds a total of 114 litres contained in two radiators, various pipe work, the engine, and the heat exchanger. Water is circulated by a water pump situated at the rear of the engine block. There is also a device at the rear of the vehicle whereby coolant can be transferred from one vehicle to another.

Ventilation

Mounted on the front of the engine, but located in the fighting compartment is a Sirocco fan which cools the main gearbox and exhaust manifolds. Air passes the cowling surrounding the gearbox casing and is pushed by the Sirocco fan into air ducts surrounding the exhaust manifolds. Hot air is then ducted into a collection box located on the rear wall of the engine compartment, from where the warm air is drawn out by the radiator cooling fans. Butterfly valves are built into the pipes between the air collection box and the fan units. These are operated by a lever on the bulkhead, and must always be set to ON.

If temperatures are low, the volume of air cooling the gearbox may be controlled by a sliding valve situated immediately below the Sirocco fan. Excess air is transferred to the engine bay to avoid low pressure in the fighting compartment and to prevent carbon monoxide being drawn out of the engine bay into the fighting compartment.

The radiator cooling fans also draw the warm air out of the engine compartment, via apertures with flaps situated on the right and left of the engine bay.

The Sirocco fan is sandwiched between the crankshaft and the propeller shaft, which in turn is connected to the main gearbox via the turret drive housing, located on the hull floor in the centre of the fighting compartment. The propeller shaft simply passes through on its way to the gearbox input flange.

Gearbox

During the early stages of the Second World War drive trains fitted to tanks differed considerably from one vehicle to another. For example, British and Russian tanks favoured the engine, gearbox, steering and drive sprockets at the rear of the

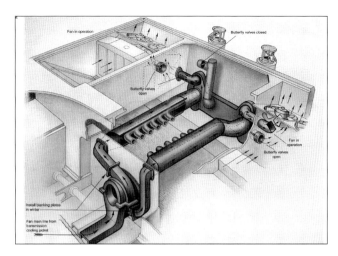

ABOVE Ventilation.

BELOW Transmission.

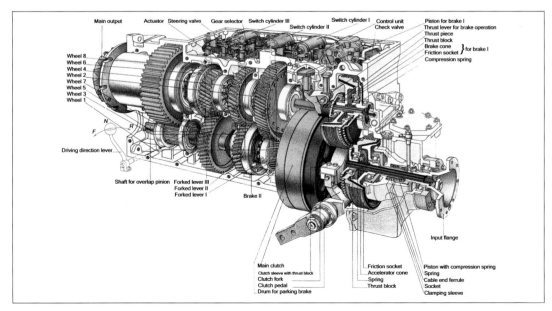

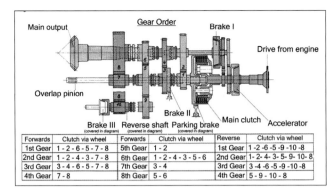

Gear Order

Main output — Brake I — Drive from engine — Overlap pinion — Brake III (covered in diagram) — Reverse shaft — Parking brake (covered in diagram) — Brake II — Main clutch — Accelerator

Forwards	Clutch via wheel	Forwards	Clutch via wheel	Reverse	Clutch via wheel
1st Gear	1 - 2 - 6 - 5 - 7 - 8	5th Gear	1 - 2	1st Gear	1 -2 -6 -5 -9 -10 -8
2nd Gear	1 - 2 - 4 - 3 - 7 - 8	6th Gear	1 - 2 - 4 - 3 - 5 - 6	2nd Gear	1- 2- 4- 3- 5- 9- 10- 8
3rd Gear	3 - 4 - 6 - 5 - 7 - 8	7th Gear	3 - 4	3rd Gear	3 -4 -6 -5 -9 -10 -8
4th Gear	7 - 8	8th Gear	5 - 6	4th Gear	5 - 9 - 10 - 8

ABOVE Plan of gearbox.

OPPOSITE TOP Gearbox oil change procedure.

OPPOSITE BOTTOM Gearbox emergency circuit.

BELOW Main clutch.

vehicle, whereas American and German tanks had sprockets, steering and gearbox sited at the front, connected to the rear mounted engine by a propeller shaft. This system made tanks more adaptable and in many cases able to increase the size of the main armament.

Why have a gearbox?

■ When Tiger 131 is in motion, where possible the engine is held at a constant 2,000rpm, which is considerably faster than the drive sprockets revolve when the vehicle is in motion.

■ Because the engine runs at a relatively high rotational speed it is also inappropriate to place the tank in or out of gear when starting from rest or stopping.

By inserting a gearbox and a driving clutch between the engine and drive sprockets, these two factors can be addressed.

Tiger I has eight forward gears and four reverse. Driving the tank in 1st gear, with engine speed at 2,000rpm, the caterpillar track drive sprockets will hardly turn at all; in fact, a pedestrian could probably walk faster.

If 8th gear were selected with the engine at 2,000rpm the tank would theoretically travel at more than 20mph, *but...* the engine does not have enough power to start from rest in 8th gear. If it were attempted the result would cause serious damage to the transmission.

Therefore, each gear has to be selected in turn so as to create a series of steps. With the engine at a constant 2,000rpm, and by selecting each gear in order, starting with a low number the rotational speed of the drive sprockets increases with each gear change, therefore increasing the overall speed of the tank.

When designing a gearbox, many factors need to be taken into consideration; for example, torque is part of the basic specification of the Tiger's engine: its power output is measured by its torque multiplied by the rotational speed of the axis which, in turn, can be measured by a dynamometer. It is vital that the right balance is achieved, when power from the engine is transmitted through the gearbox to the drive train.

Tiger 131 gearbox

Upon visual internal inspection the gearbox is much the same as a rear-wheel-drive car or lorry, insomuch that the various shafts are mounted in line with each other.

Input from the engine enters one end, and output to the steering unit is at the other end. Built into the gearbox is a semi-automatically operated clutch, to enable the vehicle to start from rest or come to a halt without stalling the engine.

A secondary shaft passes through the gearbox permanently rotating at engine speed. This shaft enables the tank to carry out a neutral turn, whereby one track drives in a

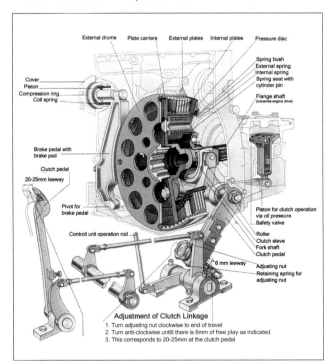

External drums — Plate carriers — External plates — Internal plates — Pressure disc

Cover — Piston — Compression ring — Coil spring

Spring bush — External spring — Internal spring — Spring seat with cylinder pin

Flange shaft (transmits engine drive)

Brake pedal with brake pad — Clutch pedal 20-25mm leeway

Pivot for brake pedal

Controll unit operation rod

Piston for clutch operation via oil pressure — Safety valve — Roller — Clutch sleve — Fork shaft — Clutch pedal — 6 mm leeway — Adjusting nut — Retaining spring for adjusting nut

Adjustment of Clutch Linkage
1. Turn adjusting nut clockwise to end of travel
2. Turn anti-clockwise untill there is 6mm of free play as indicated
3. This corresponds to 20-25mm at the clutch pedal

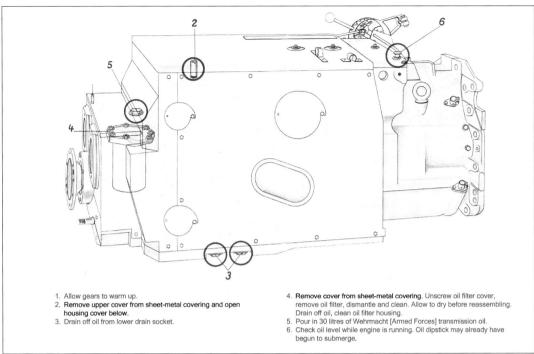

1. Allow gears to warm up.
2. Remove upper cover from sheet-metal covering and open housing cover below.
3. Drain off oil from lower drain socket.

4. Remove cover from sheet-metal covering. Unscrew oil filter cover, remove oil filter, dismantle and clean. Allow to dry before reassembling. Drain off oil, clean oil filter housing.
5. Pour in 30 litres of Wehrmacht [Armed Forces] transmission oil.
6. Check oil level while engine is running. Oil dipstick may already have begun to submerge.

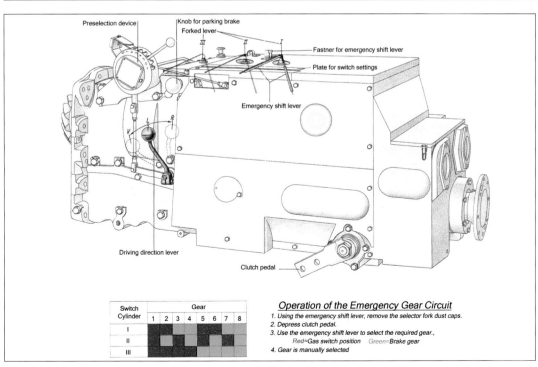

Preselection device
Knob for parking brake
Forked lever
Fastner for emergency shift lever
Plate for switch settings
Emergency shift lever
Driving direction lever
Clutch pedal

Switch Cylinder	Gear							
	1	2	3	4	5	6	7	8
I								
II								
III								

Operation of the Emergency Gear Circuit

1. Using the emergency shift lever, remove the selector fork dust caps.
2. Depress clutch pedal.
3. Use the emergency shift lever to select the required gear.,
   Red=Gas switch position   Green=Brake gear
4. Gear is manually selected

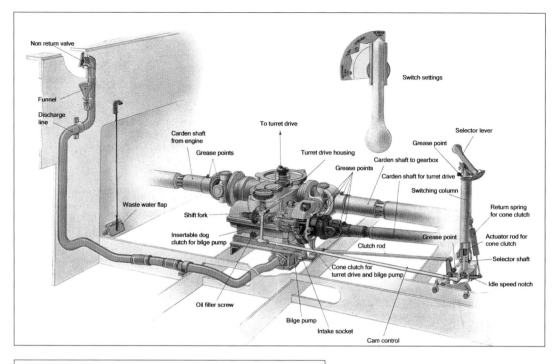

Non return valve

Funnel

Discharge line

Waste water flap

Carden shaft from engine

Grease points

Shift fork

Insertable dog clutch for bilge pump

To turret drive

Turret drive housing

Grease points

Clutch rod

Cone clutch for turret drive and bilge pump

Switch settings

Selector lever

Grease point

Carden shaft to gearbox

Carden shaft for turret drive

Switching column

Return spring for cone clutch

Actuator rod for cone clutch

Selector shaft

Grease point

Idle speed notch

Oil filler screw

Bilge pump

Intake socket

Cam control

ABOVE Turret actuator and bilge system.

LEFT Plan of steering gear.

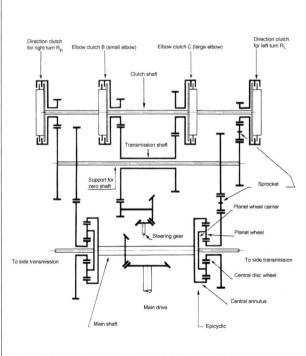

Direction clutch for right turn R_R

Elbow clutch B (small elbow)

Clutch shaft

Elbow clutch C (large elbow)

Direction clutch for left turn R_L

Transmission shaft

Support for zero shaft

Sprocket

Planet wheel carrier

Planet wheel

Steering gear

To side transmission

To side transmission

Central disc wheel

Main shaft

Main drive

Central annulus

Epicyclic

forward direction, and the other in reverse, allowing the tank to turn on its axis.

In order to carry out this manoeuvre, the tank is placed in neutral gear and the engine revolutions set to a fast tick over. By turning the steering wheel to about 45 degrees a neutral turn is achieved. Experience has shown us that the tank has to be on a hard level surface in order to limit strain on the transmission.

The advanced design of the Olvar pre-select gearbox makes it easy for the driver by using a hydraulically operated gear change. The driver selects the appropriate gear, and simply pushes the lever into the selector gate with his thumb, the remainder of the gear change is then carried out automatically. Unfortunately it isn't quite that easy. Whatever the conditions the driver must keep the engine at a minimum of 1,900rpm, otherwise there is insufficient oil pressure within the gearbox to change gear.

Turret traverse is achieved from a power take-off from the rear of the gearbox. A drive shaft connects to a small gearbox and bilge pump, via a cone clutch underneath the turret floor.

In the centre of the hull to the right of the driver's shoulder is a lever. There are four positions:

- Position 1   OFF.
- Position 2   (anti-clockwise) Bilge pump.
- Position 3   Bilge pump and turret traverse activated.
- Position 4   Turret traverse activated.

Steering

At the output end of the gearbox is the steering assembly, which is driven by bevel gears from the main gearbox. There are four multi-plate clutches in the steering unit. By operating different clutches, several activities are achievable.

- When turning the steering wheel from centre to approximately 30 degrees the tank will carry out a shallow turn.
- Above 30 degrees a slight resistance is felt on the steering wheel and the tank will carry out a much tighter turn.
- With the forward/reverse lever in neutral, as the steering wheel is turned the tank will commence a neutral turn.

The steering box can be removed from the gearbox as one unit.

Final drive

Two drive shafts (one right and one left) link the steering unit to the final drives via disc brakes. The disc brakes have two functions:

- Main brakes operated by the foot pedal.
- Emergency steering operated by levers.

The final drives are typically complex insomuch as they have a straight spur reduction gear plus an epicyclic giving an overall reduction of 10.55 to 1. In other words the output shafts from the steering unit turn 10.55 times to 1 turn of the sprockets driving the tracks, which weigh just under 6 tons in total.

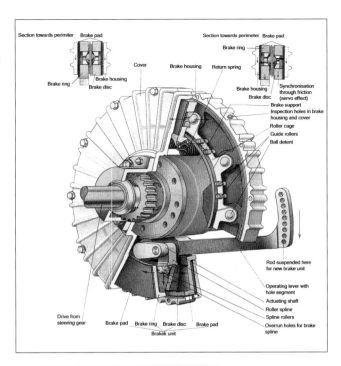

ABOVE Motor and steering brake.

Capacities

Coolant

Entire cooling system	114 litres

Armed Forces engine oil

Engine	30 litres
Air filter x 2 (HL230)	2 litres each

Armed Forces transmission oil E8

Gearbox and steering unit	30 litres
Final drives	6 litres each
Turret drive housing	4 litres
Fan drive gearbox x 2	3 litres each
Turret drive gearboxes	25 litres each

Shock absorber oil TL 6027 (violet)

Shock absorbers	1 litre each

Petrol

Total capacity	534 litres
Reserve total	140 litres

Greasing points

All greasing points on the vehicle are marked with red paint.

Lubrication and maintenance plan

Service intervals

Every 250km, 500km, 1,000km 2,000km, 5,000km.

Chapter Seven

Firepower

David Willey

Of the three key features that make up a tank – mobility, protection and firepower – it can easily be overlooked that firepower is the paramount attribute. The other two features are there to support the crew in using the firepower to get the projectile to the target. With the Tiger, its firepower came from the famed 8.8cm *KwK* 56 gun. It was this feature that clearly differentiated the tank from its immediate contemporaries and propelled tank gunnery forward in the Second World War.

OPPOSITE Admiration or fear? An Army officer ponders the 8.8cm on Tiger 131 after its capture.

Tiger gunnery

In the 1930s tanks had been designed to carry different types of guns to deal with different battlefield situations and to meet their intended role on those battlefields. Most European countries had classed tanks into three main types, Light tanks for reconnaissance, Medium tanks for exploitation and Heavy tanks for infantry support.

It was generally assumed reconnaissance vehicles needed little in the way of firepower as their role was seeking out the location of the enemy and weaponry was there more for moral support to the crew, not for serious, sustained fighting. Medium tanks were to be armed with weaponry to take on other tanks likely to be met in armoured counterstrokes and Heavy tanks needed weaponry to assist the infantry in attacking prepared positions such as trenches and bunkers. High-explosive rounds would be required in these circumstances to create blast effect along with machine-gun fire from secondary armament. In British infantry tanks,

BELOW The Flak 8.8cm in the anti-tank role. This captured example in its extemporised gun pit is inspected by Eighth Army troops.

anti-tank weapons were provided as the main armament, supposedly to fight off any counter-attacking tanks.

As the Second World War progressed, the idea of certain tanks for certain circumstances lost ground as the inevitable happened of not having the right type of tank available in the right place at the right time. A good example is the failure of the American theory of the 'tank destroyer' – lightly armoured vehicles with powerful anti-tank guns to speed to counter armoured thrusts. The expectation they would always be in a position to ambush oncoming armour and not have to engage in developing combat was cruelly exposed in the Tunisian campaign.

The doctrine for the standard US tank (the M4 Sherman) was to actually avoid tank on tank engagements – that was not its proposed purpose. The progression towards a universal tank, with a powerful gun, good mobility and adequate armour led ultimately to the Main Battle Tank of the post-war world.

The Tiger can appear to prefigure the Main Battle Tank but it was created initially as a heavy, breakthrough tank. It was also considered to be a prime tank killing weapon. The wartime circumstances of the German forces when the Tiger came into service led it to become – with a few exceptions - a specialist defence tank, rushing to counter armoured threats in both the East and West. Its gun was an outstanding attribute whether in defence or attack.

Measuring guns

Early guns were measured by the weight of shot fired – a 6-pounder gun, for example, fired shot weighing 6 pounds. Barrel size across the diameter of the exit hole or bore point of the gun was another measurement method. This sizing was also known as the calibre of a gun. Confusingly in Britain, the two systems were used simultaneously to measure gun sizes and to add to the confusion some countries used imperial and metric measuring systems at the same time.

The calibre of a gun is not the only influence on the firepower it can unleash. There is the energy behind the round when it leaves the barrel, the speed often being measured as muzzle velocity. The faster the projectile leaves the barrel the more strength there is behind it and the more penetrative power the missile will have. This force is known as kinetic energy and throughout much of the war, increasing the kinetic energy was the key aim of gun designers so that more armour plate could be penetrated.

Increasing the calibre of the gun allowed for larger amounts of force to be applied to the base of the projectilc being fired, but improvements could also be achieved by increasing the size of the charge in the propelling cartridge and lengthening the barrel of the gun. A longer barrel could also assist with accuracy.

The length of the barrel was measured by dividing its length by its diameter. For example, a 1.5m-long barrel with a 5cm bore (calibre) would be described as a 5cm L/30 gun. In Germany, guns designed for tanks were designated *Kampfwagenkanonen* (tank gun) which was abbreviated to *KwK*. The Tiger was fitted with an 8.8cm *KwK* 36 L/56 gun. The Tiger gun was based on the famous Flak 36 anti-aircraft gun – hence *KwK* 36.

An impressive pedigree

The Flak 36 had in turn antecedents in the 8.8cm *SK* Flak gun designed and built by Krupp in the First World War. The 8.8cm size of shell was considered the optimum size by the German authorities in the 1930s to fire at aircraft at medium and high altitude (then defined as from 500 to 6,000m). Designed as a high-velocity gun to fire at aircraft targets it was also adapted to fire at ground targets. This led to the issuing of an armour-piercing round, the *Panzergranate* or *Pzgr*. It was envisaged the gun might be used to knock out bunkers and pill boxes in the Maginot Line and was issued with sights (the *Flakzielfernrohr* 20E) so that it could aim at ground targets up to 9,400m away.

The gun was fitted to a number of armoured half-tracked prime movers or *Zugkraftwagen* for the campaign in Poland and later France where they proved very successful. The towed anti-aircraft gun was also used in the anti-tank role in France but statistics showed this was only a minor and infrequent part of its operational service. For example, from 9 to 26 May 1940 the *Flakregiment* 102 attached to Guderian's XIX Army Corps reported it had destroyed 208 aircraft, 13 tanks, 10 bunkers and 13 machine gun nests. From the few reports available, it was possible to show on average that it took 11 rounds to claim one tank kill from a Flak 36.

The Führer's wish

It was Hitler who insisted on the use of the 8.8cm gun on a tank chassis. Hitler held his tank symposium at the Berghof on 18 February 1941 where he demanded the installation of the 50mm and 75mm guns of higher velocity on the Panzer III and Panzer IV tanks. At the same meeting, he overruled objections from the military regarding the release of soldiers to return to factory work – making tanks was a priority. On 26 May a second symposium was held in which he further propounded his tank theories. Again, sweeping aside reservations from both the Army and industry, he insisted on the installation of the 8.8cm gun should the proposed 7.5cm Waffe 0725 tapered bore gun not be practical on the proposed Henschel VK 36.01 heavy tank project. The 7.5cm weapon was dependent on sufficient tungsten being available for the specialised ammunition that it would fire.

ABOVE The stencilled *Braun Ark* was a reminder of the fluid the gun hydraulics contained – a mixture of brown buffer fluid (*Bremsflüssigkeit braun*) and Arctic fluid (*Bremsflüssigkeit Arktisch*). The gun could recoil back to 620mm, but a marker on this gauge at 580mm warned '*Feuerpause*' or 'Stop Firing'.

At the meeting, Hitler also insisted on 100mm of frontal armour and 60mm of side armour on the new tank project for which Porsche and Henschel would now compete. In July, the lack of sufficient tungsten meant the 8.8cm gun would have to be fitted and so *Wa Pruf* 6 ordered Henschel to use the Krupp-designed turret to speed the design process. This turret therefore became the same on both the competing Henschel and Porsche Tiger tank designs – the VK 45.01 (H) versus the VK 45.01 (P).

As we are aware, despite expectation to the contrary, the Henschel design won the competition and contract. It is of interest to remember that the turret that we are so familiar with on the Tiger was originally designed for a very different tank.

Fitting the 8.8cm to a tank

The Flak gun had to go through considerable changes to create a weapon suitable to fit in a tank turret. The recoil of the Flak gun on firing was considerable so a hydraulic buffer was fitted inside the turret to the right of the gun and a hydro-pneumatic recuperator to the left. To lessen the recoil the gun was given a double baffle muzzle brake. Muzzle brakes work by capturing the expanding gases that shoot out the barrel after the round has been fired. They push the gun barrel forward, away from the tank and in so doing, counter much of the recoil action of the barrel. Some muzzle brakes can reduce the recoil on a gun by as much as two-thirds. The muzzle brake on the Tiger was replaced by a lighter version during the production run. The lighter version had been designed for the new longer and more powerful 8.8cm *KwK* 43 L/71 gun designed for the Tiger Ausf B or King Tiger.

The *Tigerfibel* states the muzzle brake on the Tiger reduced the recoil by 70 per cent and warned not to fire the gun if it was shot off or damaged. The hydraulic buffer or brake on the right of the gun was filled with oil and acted like a shock absorber. It took 25 per cent of the

RIGHT The muzzle brake. Later Tigers had a smaller version designed for the longer-calibre gun fitted to the Tiger II. This is one of the many drawings made for the report on the tank at Chobham.

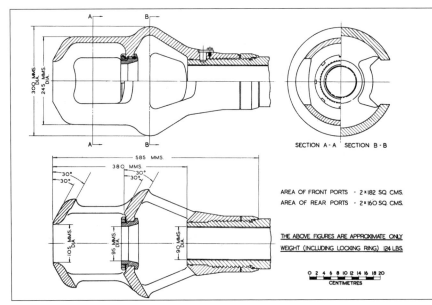

recoil, the remaining 5 per cent being taken by the hydro-pneumatic recuperator on the right of the barrel. This acted to push the gun back into the firing position.

A total of 1,514 guns were assembled and accepted by the *Heereswaffenamt* (Army Weapons Office – or HWA) inspectors from the two main assembly companies DHHV (the snappily named Dortmund-Hörder Hüttenverein AG) and Wolf Buchau. Each barrel cost 18,000 Reichsmarks. Tiger 131's gun was made by DHHV and is stamped with their identification code 'amp'. The gun barrel and breech have the serial number 'R179' and the year date '1942' stamped onto them.

As 1,354 Tiger Is were produced leaving only 160 'spare' barrels it seems unlikely that many tanks had their barrels changed during service life. The barrel life was estimated at 6,000 rounds, but this was dependent on the type of round being fired.

The barrel was rifled to spin the round for more accurate flight. The rifling grooves were 1.5mm deep and 5mm apart when first manufactured, and twisted right-handed along the barrel. These would wear during firing and make the gun slightly less accurate over time.

Holding the gun

The gun was held in place by the massive cast gun mantlet. This piece of metal was secured to the main turret by the trunnions that extended out of the side of the turret to form the two front lifting eyes for the turret (a third stud was added at the rear of the turret – hidden from view by the turret bin). The gun was not well balanced in the turret – more weight was forward of the trunnions – so a strong adjustable spring was housed inside the turret to the right of the loader's position. This through a series of levers pushed down on the inner mantlet to help balance the gun.

When not in use, a travel lock suspended from the turret ceiling hooked onto studs on the side of the breechblock. Travel locks secured the gun from unnecessary strain and possible movement of the barrel when not in combat. The lock design changed during the production run of the Tiger as crews had complained of the time it took to release and bring the gun into action.

It must be remembered that the Tiger had

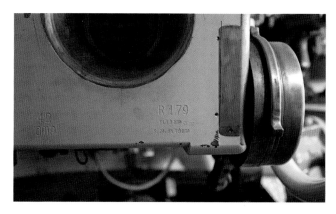

ABOVE Markings on the breech of the 8.8cm gun. The year of manufacture '42' = 1942, and the three-letter code for the maker 'amp' = *Dortmund Hoerder Huttneverein* (DHHV), appears on the lower left, while the gun's serial number 'R (for *Rohr* or piece) 179' appears on the right with 'FL 79 amp' = *Fertig Lieferant* 79 (contract number 79 with DHHV) and to the right of this two Waffen amp acceptance stamps. The 'S: M: 79 FL amp' is presumably another contract marking.

BELOW Looking back from the breech. The bar in the foreground stopped the loader from accidently getting behind the recoiling breech when firing. The leather stop pad cushioned ejected empty propellant cases. The two boxes mounted on the turret wall contained a set of headphones and throat microphone. The black tin sheet on the far side of the breech was to protect the commander from any back-flash coming from the breech opening after firing.

to stop to fire accurately. Firing on the move
with an unstabilised gun would simply waste
ammunition. The weight of the vehicle on the
wide tracks gave a solid base for accurate firing
of the gun and film of the tank firing shows the
recoil of the barrel but little movement of the
mass of the hull.

Traverse and elevation

To traverse the gun, the gunner used a pivoting
foot pedal. The pedal related to two traverse
speeds that could be selected with a gear lever.
By pushing forward on the pedal the turret would
traverse to the right, and by pushing down on
the rear it would turn left. The turret drive was
powered by a hydraulic system that took power
from the main transmission positioned under the
turret floor.

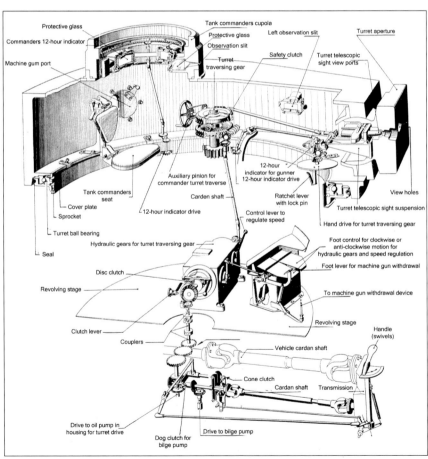

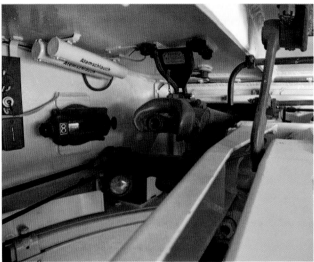

At low engine speed it could take up to 360 seconds to complete a full turret rotation. At high speed the turret could complete a full turn in 60 seconds. This system would point the gun in the general direction of the target. Fine adjustments were then made with the gunner's hand and elevation wheels. The gun could depress to –8 degrees and elevate to +15 degrees from the horizontal.

Should the hydraulic system fail, the commander or gunner could turn the turret by the much slower hand traverse method. It would take 720 turns of the hand wheel to fully rotate the turret through 360 degrees – exhausting work.

Aiming

The gun was aimed by sighting through the *Turmzeilfernrohr* 9b binocular gun sight. This was fixed to a roof bracket and the front mantlet. It has a pivoted design to allow the eyepieces and rear of the sight to remain stationary as the front, secured in the mantlet changed angle. The sights were made by Leitz – 1,253 were made before production for the Tiger switched to the monocular TZF 9c version in March 1944.

The sight had a 2.5x magnification with a field of view of 25 degrees. This would equate to seeing 444m of vision of terrain at a range of 1,000m. The graticule pattern on the right scope consisted of a number of inverted 'V'

ABOVE LEFT Beneath a gas mask, a *Gurtsack* hangs from one of the metal brackets that line the turret wall.

ABOVE Looking forward across the breech toward the gunner's sight. The dial at lower left is an azimuth indicator that reads in parallel with one in the commander's cupola. The vision port is marked '8' – one of the 14 listed items that had to be sealed before the tank could submerge. The two tubes marked '*Atemschlauch*' contained breathing tubes to attach to the standard issue gas masks. This would allow the crew to breathe filtered air, but still use scopes and vision ports without the full face mask in place.

shapes to assist the gunner in judging the distance to the target and therefore increasing the chance of a hit. The scale on the right of this telescope went from 0 to 4,000m and on the left of the same scope a range for the machine gun went from 0 to 1,200m.

Each inverted 'V' was 4 mils apart – this gap was known as a *Strich* (or mark) and tables were issued so the gunner knew the

LEFT The safety and fire lock for the loader on the side of the breech. The gun could not be fired until the loader had set the lock to 'fire'.

size of a given tank in relation to the *Strich* measurements. He could therefore judge the distance to the target tank. The *Strich* also assisted in aiming ahead (or 'leading') a moving tank target. The armour-piercing round travelled at 810m a second so a target at 1,000m away moving laterally to the tank would need judgement as to the speed it was travelling before choosing the appropriate *Strich* in the sight. Only then could the gun be aimed and fired.

As part of the *Tigerfibel*, charts were issued to assist in recognising target tanks, in judging the distance to the target and how close to the target the gunner needed to be to penetrate the armour. Considerable space is given in the *Tigerfibel* providing advice, hints and practical instruction on the use of the sights and estimating range and direction of targets.

ABOVE Looking forward from the loader's position. The hand wheel (to the right of the eye pad) gives fine adjustment to the traverse. On the bracket above the sight a notice warns that the sight must be secured in position.

BELOW Examples of information sheets on target tanks issued with the *Tigerfibel*.

Ammunition

It has already been mentioned that the Tiger's main gun was the outstanding feature at the time it went into production. The gun is present to deliver the projectile onto the target, and by doing so, achieve the aim of the whole vehicle.

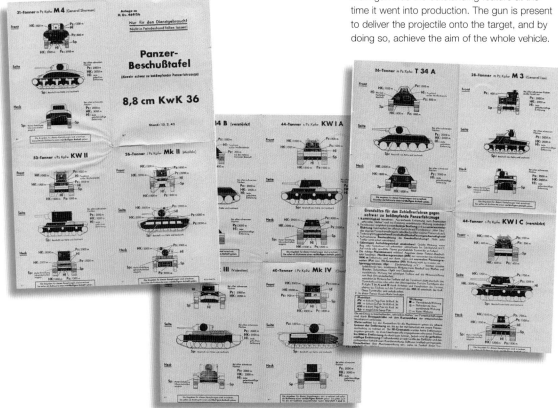

With this in mind it is worth paying attention to the types of ammunition the tank fired and their effects on the target.

The Tiger carried 92 rounds for the main gun. This was usually divided equally between high-explosive and armour-piercing rounds. These two main types of round were designed for very different purposes. Both came as single-piece ammunition, with the propellant and the projectile attached.

High-explosive rounds

The *Sprenggranaten* (*Sprgr*) or high-explosive shell weighed 9kg, of which .70kg was the explosive charge, made of amatol. The round could be supplied with a timed fuse, the *ZtZ* S/30, which could be set to explode between 2 and 35 seconds, or an impact fuse, the *AZ* 23/28, which exploded fractionally after impact. The round was fired at targets such as infantry, buildings, and artillery positions where the blast

ABOVE LEFT 8.8cm rounds fired by the Tiger. From left to right – green: High Explosive (land service); yellow: High Explosive (air service but often used in tanks); and black: Armour-Piercing.

ABOVE A view of the turret floor showing the hinged access plate lifted to reveal the ammunition stowage beneath.

damage from the shattering metal casing would cause most effect. The head of the round was painted yellow.

The blast effect of the High Explosive round sent lethal shrapnel 20m sideways to impact and 10m in front from the point of impact so pin point accuracy was not always required to achieve the desired effect.

Armour-piercing rounds

The *Panzergranate* (*Pzgr*) or armour-piercing round weighed 9.6kg and included a bursting charge of 160g. In 1942, the size of the explosive in the projectile was reduced to 59g.

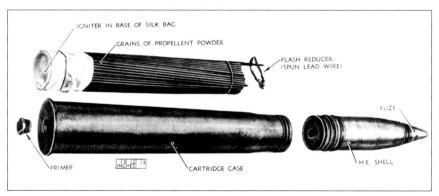

LEFT An illustration from a wartime ammunition manual of the *Sprenggranaten* (high-explosive) round, showing the projectile and propellant.

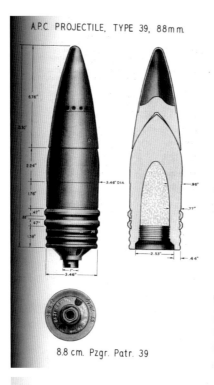

RIGHT The Panzergranate (Pzgr) 39 – the standard armour-piercing round.

A.P.C. PROJECTILE, TYPE 39, 88mm.

6.76"

3.10"

2.24"

3.46" DIA

1.76"

.96"

1.47"

22°

.47"

.77"

1.38"

2.53"

.44"

3.46"

8.8 cm. Pzgr. Patr. 39

A.P. PROJECTILE WITH TUNGSTEN CARBIDE CORE (TYPE 40) 88mm.

BALLISTIC CAP

AIR SPACE

PLASTIC

12.75"

TUNGSTEN CARBIDE CORE

3.43" Dia

AIR SPACE

.20"

.43"

1.16"

TRACER

3.46"

8.8 cm. Pzgr. 40

RIGHT The Panzergranate (Pzgr) 40 carried an inner core of tungsten. During the Second World War, Germany's reserves of Tungsten were diminishing and so proportionate to overall projectile production very few Pzgr 40 rounds were made.

This round – known as the *Pzgr 39* – had an increased overall weight of 10.2kg.

Initially designed to smash through armoured positions such as pill boxes on the Maginot Line, the armour-piercing round was developed to become a fearsome anti-tank round. The idea behind most armour-piecing rounds is to project a solid lump of hardened metal with massive kinetic energy at a target. By firing the shot as hard and as fast as possible, the sheer momentum will smash the round into and penetrate the target, for example the armour of an enemy tank.

Most solid shot breaks up into hot fragments when having penetrated a tank. The fact that this could cause crew injury, major internal damage or fires was not enough for the German designers, so they included a bursting charge within the round (technically turning it from 'shot' to 'shell').

Hence in the language of ammunition, the *Pzgr 39* was an Armour-Piercing, Capped, Ballistic Capped (APCBC) round.

Composite rounds

The *Pzgr 40* round worked on a similar principle to the usual armour-piercing round but it housed an internal metal core made of tungsten. Tungsten is twice the weight of steel and is incredibly hard. The aim of a composite shot was to give the full force of the calibre of the gun concentrated behind the carrying body. This body surrounded the hard tungsten core. The overall weight of the round was less than the usual solid armour-piercing round. This enabled more kinetic energy behind the shot and only the tungsten core would penetrate the target. But Germany had limited access to the metal. When available for issue, *Pzgr 40* rounds were limited for use against later, more heavily armoured Russian tanks such as the KV or Joseph Stalin series.

Armour-Piercing impact

Most armour piercing rounds fired at a tank will not meet a flat 90-degree surface. Tanks are complex mechanisms with armour protection at varying angles and thicknesses and other complexities like stowage, wheels, and spare track etc on the outside. This will rarely allow a round a simple meeting with the armoured surface of the vehicle.

Should a round hit the target at 90 degrees, if it has enough energy, it carries on into the plate causing metal to deform, making a collar or 'petalling'. Further penetration can cause deforming of the metal into a bulge of the armour plate ahead of the round. The armour plate can then fail by 'plugging' – that is, the armour ahead of the round comes out as a core or plug, or it can break off as a disc (or a combination of both). At any point during this process if the round looses enough kinetic energy it may fail to penetrate the target.

On impact on angles over 20 to 30 degrees rounds, tend to follow an S-shaped path into the armour. The bulging or plugging of metal ahead of the round creates a weakness and the shot follows this line of least resistance, causing the round to change angle during its penetration of the metal.

On hitting angles over 60 degrees, there is a much higher chance that the round will simply ricochet off the surface metal.

There are a number of ways shot hitting an armoured target can break up or fail. If the shot is too soft it can compress and spread, causing 'barrelling', and lose the energy to penetrate. It can also bend on impact or simply shatter.

To stop shattering a cap of metal was designed to fit over the end of the shot. This serves to cushion the round on impact before it disintegrates and allows the main part of the shot to start penetrating the armour of the target. A further refinement was later made to this cap to act as a swivel point so on hitting armour plate at an angle, the cap would turn away as if ricocheting off and in doing so would force the round behind to turn more directly to attack the armour surface at 90 degrees.

The cap added to prevent shattering had a poor ballistic shape. It created a blunt end that slowed the round in flight. To counter this, a thin ballistic cap was fitted over the round to create a more aerodynamic shape. This ballistic cap would simply collapse on contact with the target.

Hollow-charge rounds

The firing of kinetic energy rounds was not the only way to defeat armour and knock out

a tank. The development of HEAT, or High Explosive Anti-Tank rounds, had progressed before the Second World War. The HEAT round uses something called the Munroe effect. If explosive has a wedge shape on its surface, when it is exploded against a surface the detonation wave is concentrated and more penetration of the surface is achieved. By moving the explosive away from the surface (the stand-off distance) the effect is further increased. If the explosive is formed into a hollow cone shape lined with metal, when exploded about 20 per cent of the metal forms a jet of super-heated metal that creates an intense kinetic energy point (an effect of about 200 tons pressure per square inch). This forces its way through armour plate followed by the remaining 80 per cent of the cone metal in a rod or plug shape. If it penetrates the armour, spall (metal parts) can cause damage to the tank interior and pressure waves injure the crews' eardrums or cause temporary blindness.

The amount of explosive required in an effective HEAT round is relatively small and its effect and penetrative powers are considerable. There is a relation between the diameter of the cone of explosive and the thickness of metal it can penetrate – usually this means armour plate, three or four times the diameter of the explosive cone can be penetrated.

The *Gr 39 HL* was a hollow charge HEAT round developed for the 8.8cm gun but its penetration was not as great as the *Pzgr* 39 (only 90mm at all ranges) and it was less accurate. Its advantage was that it could be used as a high-explosive round as well as armour-piercing.

The hollow-charge round is perhaps best remembered in German wartime service in the guise of the *Panzerfaust*, the mass-produced single-shot weapon that could penetrate 200mm of armour.

Available production figures for the 8.8cm round show the following;

	1942	1943	1944
Sprgr (*Sprenggranaten*)	14,100	1,392,000	459,400
Pzgr 39 (*Panzergranate*)	21,200	324,800	394,000
Pzgr 40 (*Panzergranate*)	8,000	8,900	None

Penetration

The penetrative ability behind the 8.8cm *KwK* 36 L/56 gun was tested by the German authorities. Their measurements were reached by firing against and penetrating different thicknesses of steel plate set at 30-degree angles.

The *Pzgr* 39 projectile could defeat 10cm of plate at 1,000m and 8.4cm of plate at 2,000m. It is salutary to remember the Sherman M4A1 had 7.4cm as its thickest armour on the gun shield.

Accuracy

The German weapons authority also extensively tested the accuracy of the 8.8cm *KwK* 36 L/56 gun. The target was 2.5m wide and 2m high and firing was conducted at known distances; for example the *Pzgr* 39 round could be fired with 100 per cent accuracy at the target 1,000m away, dropping to 87 per cent at 2,000m and 53 per cent at 3,000m. However, these impressive figures have to be qualified as being taken in a controlled 'test' environment. With variations added due to barrel wear, ammunition quality and human error the percentage accuracy drops considerably at further distances and of course this would change again in combat where additional stresses such as terrain, atmospherics and the pressures of combat come into play.

The number of rounds therefore required to hit a target tank is still hard to define from reports and first-hand accounts, but the gun undoubtedly gave the Tiger the advantage on the battlefield. It could successfully deal with most opposition tanks at ranges beyond that with which they could return accurate or effective fire.

Stowage of ammunition

The 92 rounds were stowed in groups of 16 in the side panniers built out over the tracks. They were stowed alternately nose and tail forward. Further rounds were stowed in a reserve bin beside the driver and six under a hatch in the turret floor. The rounds sat on pivoted spring loaded rests which could be moved out of the way to access the lower racks. The bins were covered by folding metal sheets but not in

RIGHT One of the side ammunition panniers is opened to reveal the racking inside.

BELOW A cutting from a wartime German newspaper. Loosely translated it says: 'Loading the Tiger with ammunition. In the counter-attack at Bjelgorod [Battle of Kursk] these tanks once again showed their supremacy.'

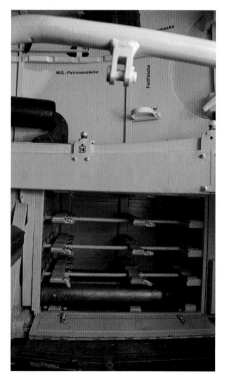

Munition für den „Tiger". Diese Panzer haben bei dem Gegenangriff im Raum von Bjelgorod erneut ihre gewaltige Überlegenheit bewiesen

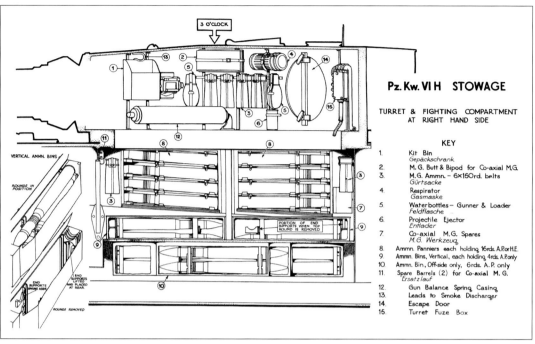

Pz. Kw. VI H STOWAGE

TURRET & FIGHTING COMPARTMENT
AT RIGHT HAND SIDE

KEY

1. Kit Bin
    Gepäckschrank
2. M.G. Butt & Bipod for Co-axial M.G.
3. M.G. Ammn. – 6×150rd. belts
    Gürtsacke
4. Respirator
    Gasmaske
5. Waterbottles – Gunner & Loader
    Feldflasche
6. Projectile Ejector
    Entlader
7. Co-axial M.G. Spares
    M.G. Werkzeug
8. Ammn. Panniers each holding 16rds. A.P. or H.E.
9. Ammn. Bins, Vertical, each holding 4rds. A.P. only
10. Ammn. Bin, Off-side only, 6rds. A.P. only
11. Spare Barrels (2) for Co-axial M.G.
    Ersatzlauf
12. Gun Balance Spring Casing
13. Leads to Smoke Discharger
14. Escape Door
15. Turret Fuze Box

VERTICAL AMMN. BINS

ROUNDS IN POSITION

END SUPPORTS LIFTED AND PLACED AT REAR.

END SUPPORTS SPRING ASIDE.

ROUNDS REMOVED

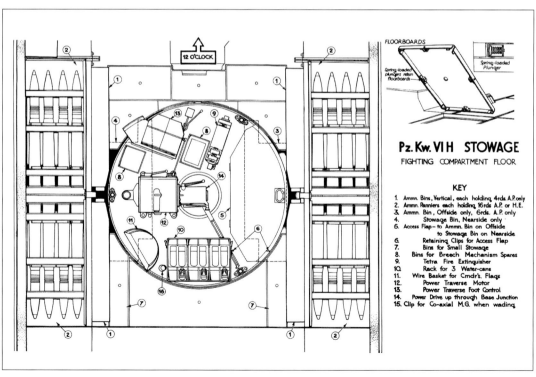

FLOORBOARDS

Spring-loaded plungers retain floorboards

Spring-loaded Plunger

Pz. Kw. VI H STOWAGE

FIGHTING COMPARTMENT FLOOR

KEY

1. Ammn. Bins, Vertical, each holding 4rds. A.P. only
2. Ammn. Panniers each holding 16rds. A.P. or H.E.
3. Ammn. Bin, Offside only, 6rds. A.P. only
4. Stowage Bin, Nearside only
5. Access Flap – to Ammn. Bin on Offside
    to Stowage Bin on Nearside
6. Retaining Clips for Access Flap
7. Bins for Small Stowage
8. Bins for Breech Mechanism Spares
9. Tetra Fire Extinguisher
10. Rack for 3 Water-cans
11. Wire Basket for Cmdr's. Flags
12. Power Traverse Motor
13. Power Traverse Foot Control
14. Power Drive up through Base Junction
15. Clip for Co-axial M.G. when wading

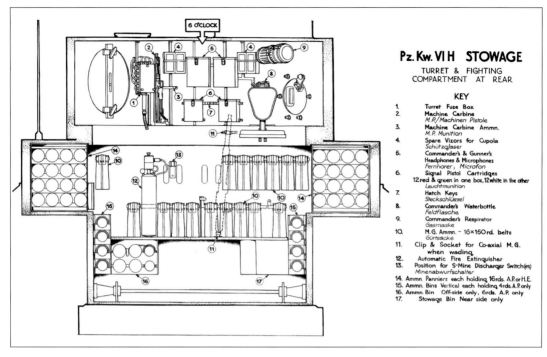

Pz.Kw.VIH STOWAGE
TURRET & FIGHTING COMPARTMENT AT REAR

KEY

1. Turret Fuze Box
2. Machine Carbine
   M.P./Machinen Pistole
3. Machine Carbine Ammn.
   M.P. Munition
4. Spare Vizors for Cupola
   Schützglaser
5. Commander's & Gunner's Headphones & Microphones
   Fernhorer, Microfon
6. Signal Pistol Cartridges
   12 red & green in one box, 12 white in the other
   Leuchtmunition
7. Hatch Keys
   Steckschlüssel
8. Commander's Waterbottle
   Feldflasche
9. Commander's Respirator
   Gasmaske
10. M.G. Ammn. – 16×150 rd. belts
    Gürtsacke
11. Clip & Socket for Co-axial M.G. when wading.
12. Automatic Fire Extinguisher
13. Position for S-Mine Discharger Switch(es)
    Minenabwurfschalter
14. Ammn. Panniers each holding 16 rds. A.P. or H.E.
15. Ammn. Bins Vertical each holding 4 rds. A.P. only
16. Ammn. Bin Off-side only, 6 rds. A.P. only
17. Stowage Bin Near side only

armoured or jacketed containers as became the norm in Western Allied tanks.

Ammunition was delivered in wooden boxes with two rounds in each container. The boxes were supposed to be returned for reuse but combat situations could prevent this. Rearming a tank was a strenuous affair for the crew with each round weighing up to 20kg. The overall ammunition load was over 1,500kg.

Secondary armament on the Tiger

The tank also carried two machine guns, both the *Maschinengewehr* 34 (MG34). This machine gun was designed to meet the German Army's requirement for a 'universal' gun – one that could be adapted for a number of roles; sustained fire (on a complex MG-Lafette quadruped), as a section weapon for carrying into action (on a simple *Zweibein* 34 bipod) and as an anti-aircraft gun (on *Dreifuss* 34 tripod). The gun was also mounted on twin anti-aircraft mounts and adapted for use in vehicles.

Despite 345,109 guns being made by the end of the war there were simply never enough. One reason for this, and the flaw with this reliable and well-respected weapon, was the complexity of manufacture. Tolerances were fine – which meant the addition of sand or mud on the battlefield could cause jams. The gun

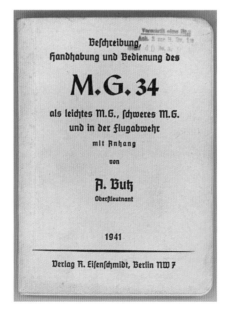

RIGHT MG34 handbook and crew notes.

Procedures for aiming and firing the gun

The tank commander, sitting at the highest point in the vehicle would select and direct the gunner who sat immediately in front of him in the turret. The commander would direct where the target was, the type of ammunition to fire and the estimated range to the target. If the driver gave a different estimate of range, the gunner would average the distance.

Training exercises give a good indication as to the procedures to be followed:

Tank Commander High explosive – load!
Loader (on completion of task) High explosive loaded, safety on.
Tank Commander Gunner, one o'clock, high explosive, anti-tank gun, 900 metres, fire when ready!
Gunner Achtung! Firing.

The gunner then reports the fall of the shot and calls the correction needed. The Commander may also call 'Over!' or 'Short!'

Further rounds are fired until the target is dealt with to the commander's satisfaction. He will then state:

Tank Commander Stop loading!
Loader Clear.

The loader rarely wore headphones in the tank in a combat situation as his role meant rapid movement and leads would restrict his actions.

The actual firing was carried out by the gunner from a trigger bar set on the inside of the elevating wheel. This sent an electrical charge to the base of the shell where a C/22 electrical primer was fitted. This differed from the C/12 percussion primer fitted to rounds fired by the standard 8.8cm Flak gun that housed a traditional firing pin. The electrical charge caused instantaneous firing. However, there were complaints from crews that should the tank be hit, the jarring action could cause the melting of fuses – fuses that were hard to replace on campaign. This led to the gun having to be fired from an emergency firing device.

ABOVE The loader's seat faced backwards in the turret and could be lifted and pivoted out of the way when in action.

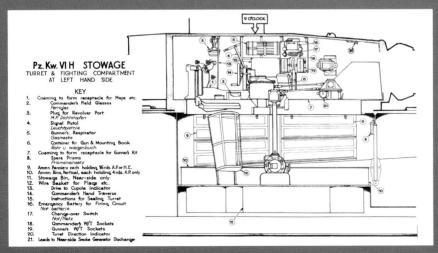

Pz. Kw. VI H STOWAGE
TURRET & FIGHTING COMPARTMENT
AT LEFT HAND SIDE

KEY

1. Coaming to form receptacle for Maps etc.
2. Commander's Field Glasses
   Fernglas
3. Plug for Revolver Port
   M.P. Dichtstopfen
4. Signal Pistol
   Leuchtpistole
5. Gunner's Respirator
   Gasmaske
6. Container for Gun & Mounting Book
   Rohr u. wiegenbuch
7. Coaming to form receptacle for Gunner's Kit
8. Spare Prisms
   Prismeneinsatz
9. Ammn. Panniers each holding 16 rds. A.P. or H.E.
10. Ammn. Bins, Vertical, each holding 4 rds. A.P. only
11. Stowage Bin, Near-side only
12. Wire Basket for Flags etc.
13. Drive to Cupola Indicator
14. Commander's Hand Traverse
15. Instructions for Sealing Turret
16. Emergency Battery for Firing Circuit
    Not batterie
17. Change-over Switch
    Not/Netz
18. Commander's W/T Sockets
19. Gunner's W/T Sockets
20. Turret Direction Indicator
21. Leads to Near-side Smoke Generator Discharger

9 O'CLOCK

ABOVE The armoured barrelled MG34 for tanks (right) is next to the standard issue weapon.

ABOVE RIGHT Staff Sergeant James Jarvis of REME took the Bakelite butt of the MG34 from Tiger 131.

RIGHT An example of a *Gurtsack* that could carry 150 rounds of belted 7.9mm ammunition. The Tiger carried 32 *Gurtsäcke*.

a range of types including incendiary, armour piercing and tracer. Tracer was often placed at regular intervals within a belt of standard ammunition so the fall of the shots could be observed. Tracer burnt out at about 1,000m.

The version of the MG34 for armoured vehicles – called the MG34 *mit Panzermantel* – had an armoured barrel covering two-thirds of its length. This was to protect the gun outside the vehicle from bullet splash and light shrapnel. The standard barrel covering was a relatively thin steel pierced covering to assist with the air cooling.

Bow machine gun

The bow machine-gunner also operated the tank radio, which sat on a bracket between his position and the driver's. The gun fitted into a *Kugelblende* 50 ball-mount. This ball mounting allowed for a 15-degree movement to the left and right of centre and a –10 to a +20 degrees vertical movement. The gun has the standard

was joined in service by the much simpler to produce MG42.

The gun had a rate of fire up to 900 rounds per minute – which was fast and gave the gun a distinctive sound as it fired ('like ripping calico'). It fired the standard 7.92mm ammunition (used in the standard German rifle), which came in

RIGHT The MG34 *mit Panzermantel* fitted in the ball mounting. The head pad to help elevate the gun was criticised in British reports as being painful to use.

LEFT The bow machine-gun position with MG34 in place. The head cap was supposed to assist the firer in manoeuvring the gun; the sight, the *KZF* 2, sits under the rubber brow pad to the left of the gun.

CENTRE LEFT The bow machine-gunner and radio operator's position. The MG34 has been removed but a *Gurtsack* hangs on its bracket and the *KZF* 2 sight is in place with a rubber eye pad fitted. In the corner more *Gurtsäcke* hang on their brackets and two standard MG34 ammo boxes sit on the pannier bracket.

BELOW The *KZF* 2 sight – or *Kugel Zielfernrohr* 2 – that was fitted to the bow machine-gun position.

BELOW The *KZF* 2 sight was only fitted to the bow MG34 machine gun; the co-axial gun was sighted with the main tank scope.

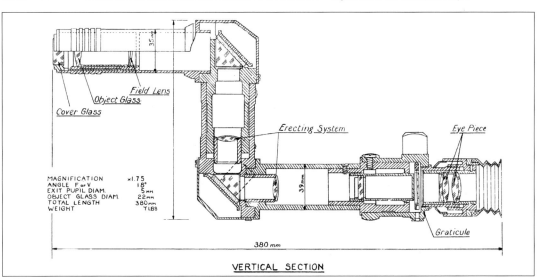

Cover Glass
Object Glass
Field Lens
Erecting System
Eye Piece
Graticule

MAGNIFICATION ×1.75
ANGLE F or V 18°
EXIT PUPIL DIAM. 5 mm
OBJECT GLASS DIAM. 22 mm
TOTAL LENGTH 380 mm
WEIGHT 7 LBS

380 mm

VERTICAL SECTION

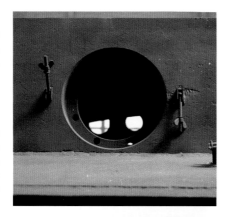

RIGHT The threaded bolts on either side of the MG port aperture were to secure the waterproof cover in place during wading.

BELOW The ball mount apart. The machined recess on the armoured plate was to locate the waterproof cover.

BELOW The breech arrangement and rear of the co-axial MG34 mount. The main gun travel lock hangs from above the ceiling and its locating stud can be seen beneath it.

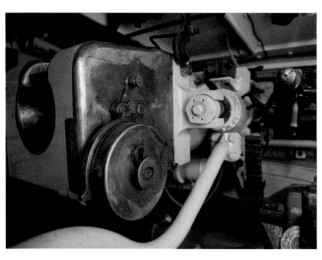

KZF 2 or (*Kugel Zielfernrohr* 2) gun sight fitted in the mount. Movement to aim the gun was by a head pan and the pistol grip. The dome shaped head pan was used to alter the elevation of the gun. The British report on Tiger 131 noted the spring to balance the gun (more weight resided at the pistol grip end) failed to relieve the pressure bearing down on the user's head.

Ammunition for the gun was stored in the standard ammunition boxes and canvas bags (called *Gurtsack*) suspended on the interior wall of the vehicle. Each bag contained a belt of 150 rounds and the Tiger had positions for 32 *Gurtsäcke* as part of the internal stowage. Two bags fitted to the rail beneath the gun, one, on the left, to hold the belt of ammunition feeding into the gun, the other, on the right, was intended to catch the spent casings after firing.

Co-axial MG34

A second MG34 was fitted as a co-axial weapon next to the main gun in the turret. This machine gun could be also fired by the gunner – a foot pedal ran through a number of linkages and levers to fire the gun. Crew reports from early combat criticised the firing system and the awkward positioning of the gun. Stoppages were hard to clear and the access to reload awkward. This gun was also fitted with the canvas bag system to provide belted rounds and a second bag was used to collect spent casings.

Sadly the MG34s fitted to the Tiger did not have their serial numbers recoded during the evaluation process. If they had they might confirm the story behind a donation to the museum in 2002. Staff Sergeant James Jarvis was part of the REME crew tasked with recovering the Tiger. Like many soldiers Jarvis was interested in acquiring souvenirs and he picked up one of the MG34 removable butts from the tank (serial number 5236). The Bakelite butt made its way back to the UK and, as a considerably older man, Mr Jarvis brought the butt along to the museum to reunite it with the tank. What other items sit unloved in attics or unknown in garages that have amazing stories attached, if only we knew them?

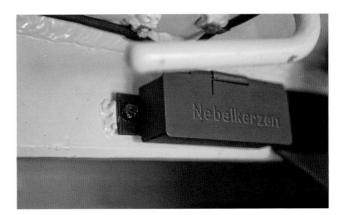

Small arms

The tank was also issued with one MP40, the sub-machine gun often erroneously called the *Schmeisser* by Allied forces. Originally designed as a compact weapon for paratroopers and vehicle crews, over a million of these guns were made during the war years. The gun was housed on the rear wall of the turret for close crew protection when away from the vehicle and in defence of the vehicle from the pistol ports in the turret sides.

Crew might carry a personal weapon such as the standard Walther P38 pistol but belts with holsters were rarely worn in combat as holsters could get in the way in the cramped tank interior. Belts were a potential snagging point during emergency evacuation of the vehicle.

It can hardly be classed as a weapon system, but the smoke candles or *Nebelwurfvorrichtungen* must be mentioned. These arrangements of stubby tubes or dischargers on either side of the turret contained smoke-generating cans that could be triggered from inside the vehicle. Two small metal boxes on the turret roof to the left and right of the commander marked *Nebelkerzen* contained three firing buttons under a sprung lid. Pressing the buttons to fire the smoke generators caused a black powder charge to project the generator out of the tube and at the same time ignite the generator to produce smoke. Combat reports stated that small-arms fire could set off the generators and wreathe the vehicle in smoke to a degree that if no wind or movement was possible – the crew could become badly affected. Fixing of the smoke candles consequently ceased in June 1943.

The one known photograph of Tiger 131 while still in German service shows the right-hand trio of dischargers having lost a tube. Subsequent to capture, this missing discharger was replaced from another vehicle before its return to the UK.

ABOVE The box is opened to reveal the firing buttons.

RIGHT Smoke dischargers are located on the turret sides.

'We were in action almost without pause for 29 months. Nobody can comprehend this. Not even us. When I reflect on it today, after 60 years – an impossible job. But we had to see it through didn't we? We couldn't just abandon the tank and go home.'

Obergefreiter Siegfried Schiller,
Schwere Panzer Abteilung 503

Fighting the Tiger

───●────────────────

Who used the Tiger, tactics
employed, communications and
the issues of mobility, reliability
and repair are all essential to our
understanding of how the tank
was used on the battlefield.

OPPOSITE Advance of Tigers from the 2nd SS Panzer Division
'*Das Reich*' near Orel during Operation 'Citadel' on the Russian
Front, 10 July 1943. The insignia on the front armour plate was
painted on all *Das Reich* vehicles, front and rear, before this
battle. *(Ullsteinbild/TopFoto 109198)*

FIGHTING THE TIGER

LEFT Generaloberst Heinz Guderian inspects a
Tiger in the spring of 1943 during a visit to the
1st SS Division 'Leibstandarte SS Adolf Hitler',
which had taken part in fighting near Kharkov.
(Ullsteinbild/TopFoto 047565)

Units that used Tigers

The Tiger was issued to specially established
heavy tank battalions, the first two, 501
and 502, were created in May 1942. They
were followed by the formation of seven
more battalions 503–510. These special
units were designated *Heerestruppen* (or
independent Army troops) and attached to
other units when deemed appropriate by higher
command. The tanks were also issued to the
elite *Grossdeutschland* Division and the three
Waffen-SS Panzer Divisions – *1st SS-Panzer-
Division 'Leibstandarte-SS Adolf Hitler'*, *2nd
SS-Panzer-Division 'Das Reich'* and *3rd
SS-Panzer-Division 'Totenkopf'*, each having an
integrated Tiger battalion.

Initially, those selected to train for Tiger units
were considered the elite of the *Panzertruppen*.
As the war progressed, the quality of recruits
and soldiers transferred to tanks diminished
and despite the German emphasis on training,
this thinning of the elite and lessening of
experience undoubtedly caused increased
numbers of breakdowns or ditched and
abandoned vehicles.

To assist the training of the crews, the
German authorities produced a small manual
known as the *Tigerfibel*. Its format – cartoons,
ditties and aide memoirs to help crews
understand and remember key operational
issues – has been ascribed to *Leutnant* Josef
von Glatter-Gotz and a similar booklet was
made for the Panther. The booklet is referred to
in a number of places in this book.

The number of Tigers issued to each
company changed during the course of the
tank's operational period. Early units fielded

LEFT Tank crews receive training on the use
of sights. The *Strich* or mark triangles from
the graticule in the sight are chalked on the
blackboard.

ABOVE Gunnery training was in redundant tank turrets installed on concrete mounts. The range was on the Baltic coast with the guns firing out to sea.

LEFT The Tiger appeared on the covers of a number of wartime magazines. *Die Panzertruppe* was bedtime reading matter for every good tank man.

three platoons (or *Zuge*) each of three Tigers. Later ten Panzer III tanks were added to the company as escort vehicles. Various combinations of the two types of tanks were experimented with. The Panzer III combination with Tigers did not last long and companies had the lighter tanks replaced by SdKfz 250 half-tracks to carry out scouting, liaison and guarding duties instead. At the same time, the number of Tigers issued to each company was increased to 14.

By March 1943 orders decreed that each heavy tank battalion should have 45 Tigers, three for the headquarters company and 14 in each of the three companies. This arrangement remained until the end of the war. Within each company, there was the combat arm; a company headquarters with two Tigers (one for the commander and one as reserve) and three platoons of four Tigers each. The wheeled or support echelon with further headquarters, medics, vehicle maintenance and combat and baggage supply trains, made up the rest of the Company.

BELOW When originally issued in late 1942 and early 1943, heavy tank battalions had SdKfz 141, the Panzer III, as escort vehicles to accompany the Tiger. Here the Tank Museum's Ausf L, also a runner, poses beside the Tiger.

Successful use of the tank on the battlefield relied not only on crew training, but also on their ability to react to changing circumstances. For this, inter-tank communication was essential.

Communications

The Tiger was issued with a *Fu* 5 radio. *Fu* stood for *Funkgerät* and the term encompassed both the transmitter and receiver. The *Fu* 5 was the standard German tank communication radio with a 10w transmitter and an ultra short wave length receiver. It operated on the standard tank frequency band of 27200–33300kHz. This radio had a usable range between 4 and 6km. However, the quality and distance of acceptable transmission was heavily influenced by the terrain and atmospheric conditions.

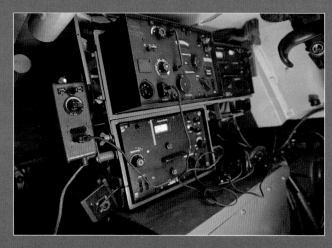

ABOVE The radios installed to the left of the bow machine-gunner's position.

RIGHT German headphones and throat microphone.

Platoon leader vehicles were fitted with two radios, the *Fu* 5 and also the *Fu* 2 (an ultra short wave length receiver).

A number of Tigers were specially converted to become *Panzerbefehlswagen* or command vehicles. These vehicles had a number of alterations to make room for extra radio equipment. The *Fu* 5 was fitted in the turret and one *Fu* 7 or *Fu* 8 in the hull. An electrical generator was also fitted and extra radio antenna mounts affixed on the hull.

Inside all vehicles there was a *Bordsprechanlage* (intercom system). The gunner, commander, driver and radio operator could all plug into this system – the loader could not as it was thought his movements in reaching and loading rounds might be inhibited by a headset or microphone. The system operated by a throat microphone and a twin earphone headset.

Radio discipline was considered vital to stop the enemy learning of any intended action or gathering useful information, hence the use of code words. Tanks and units were identified by prearranged code names, places on maps by prior agreed numbering of the locations and times given from an agreed start point – not the actual 24-hour clock. A table, or *Sprechtafel*, listing the agreed codes was changed as soon as compromised or at least every nine days.

It was also vital to keep messages brief to avoid blocking the channel for more important information. Inevitably in times of combat, stresses led to the breaking of the agreed rules and structures and sometimes it was important to give messages openly and clearly rather than risk any confusion.

The calibration of these radios was a constant issue as valve systems in a vibrating armoured vehicle caused frequencies to wander and damage to occur.

Commanders could also communicate by hand, flag and pyrotechnic signals. In combat flag signalling was rapidly dropped as exposure invited fire from the enemy but it was a useful back-up if radios failed and a basket to keep flags in was sited just behind the commander's seat. A flag was also used to indicate if the tank had broken down, usually a yellow flag with a black cross. Hand

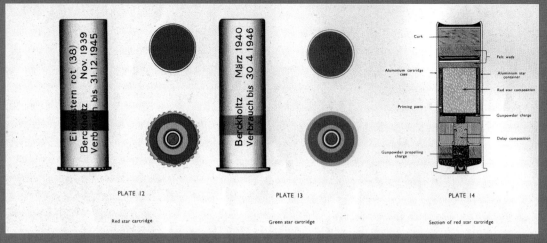

PLATE 12

PLATE 13

PLATE 14

Einzelstern rot (38) Nov. 1939
Berckholtz Verbrauch bis 31.12.1945

Berckholtz März 1940
Verbrauch bis 30 4 1946

Cork
Aluminium cartridge case
Priming paste
Gunpowder propelling charge
Felt wads
Aluminium star container
Red star composition
Gunpowder charge
Delay composition

Red star cartridge

Green star cartridge

Section of red star cartridge

signals were used in daylight hours and when radio silence needed to be observed, for example on the approach march to the start line of an operation. At night red, green or white signals were supposed to be used by signal lamp or torches but again this was often superseded by each unit creating its own system to suit the terrain.

The commander of a vehicle could also fire the *Leuchtpistole* (flare pistol). This simple weapon with a barrel calibre of 2.7cm took one round. During the war, 40 different signal cartridges and two grenade types were designed to be fired from the gun. For a tank commander, a pre-agreed coloured flare could be used to signal the launch of an attack or retreat, indicate the location or nature of an enemy position or warn off friendly aircraft.

ABOVE Agreement was made on the significance of the colouring of flares within each unit and this would be changed regularly to stop the enemy familiarising themselves with the signals. Here, from a wartime munitions identification manual, are the markings for red and green flares and the packaging artwork.

500 Signalpatronen Einzelstern, rot

Auftrags-Nr.:
Lieferfirma:
Angefertigt
Verbrauchszeit bis
Ratt.

LEFT '*500 Signalpatronen, Einzelstern, rot*' – flare ammunition packaging label.

LEFT The *Leuchtpistole* or 2.7cm flare pistol. Firing a single cartridge it was mainly for communications, but it could also be used for launching grenades.

RIGHT A Tiger from *Leibstandarte SS Adolf Hitler* passes a burning house in a Russian village during the fighting near Orel. *(Ullstein Bild/ TopFoto 041336)*

Tactics

BELOW An atmospheric shot of two Tigers of the 502nd Heavy Tank Battalion advancing past a knocked-out KV1 and BT7 (oval hatches when compared to the rounder T-34 hatches) on the Russian Front. *(Ullsteinbild/ TopFoto 047612)*

The initial use of Tigers in combat was a haphazard affair with no formal instruction given as to the best use and employment of this powerful new weapon system. By early 1943, publications on the use and employment of the tank were being issued and these clearly emphasise the role of the Tiger as a breakthrough and armour-destroying weapon. 'They will be attached to other Panzer units in the decisive point of the battle in order to force a decision.'

The tanks were to be used in at least platoon strength and utilise their superior concentrated firepower while being protected by strong armour. Different formations for the platoon were suggested depending on the nature of the terrain and attack, but the general tactic was to ensure at least one or two vehicles were stationary and ready to fire while the other two vehicles advanced or manoeuvred in up to 200m bounds to the next potential fire position.

The destruction of enemy armour was constantly emphasised in publications and training. 'They are especially suited for fighting against heavy enemy tank forces and must seek this battle. The destruction of enemy tanks creates the prerequisite for the successful accomplishment of the tasks assigned to our own lighter Panzers.'

From mid-1943, after the Kursk offensive on the Eastern Front, Tiger units found themselves less as breakthrough weapons and more as fire-fighting units being rushed from one combat area to another by rail to stem breakthroughs or put in spoiling attacks.

Again and again, single Tigers or groups of Tigers performed outstanding actions against numerically superior attacking armoured forces. Hans Bolter (*s.Pz.Abt.502*), Kurt Knispel

ABOVE Russian generals inspect the first captured Tiger in 1943. Red Army Deputy Commander-in-Chief Marshal Georgi Zhukov shows the tank to colleagues including Marshal Kliment Voroshilov and Marshal Nikolai Voronov. This very early production tank, from the 1st Company of the 502nd Heavy Tank Battalion, had the main stowage bin fixed to the side of the turret, not to the rear. This was a crew modification to allow easier access to the engine decks. *(Art Media/HIP/TopFoto 0212914)*

RIGHT Adversaries in retirement. The Tiger is next to a Polish-marked T-34/85. They illustrate the very different philosophies behind the German and Russian designers and their approach to warfare.

Leutnant *Alfred Rubbel, Schwere Panzer Abteilung 503*

'We became the fire brigade, always fighting in the hot spots during the day and shifting position at night. Of course, that didn't do the tanks any good and we had a lot of failures for technical reasons.

'When the Tiger went into action for the first time it was a piece of Hitler's idiocy. It was in swampy ground near Leningrad and two got stuck and the Russians got hold of them.

'Then came the big event; we had heard a lot about the Tiger but not yet seen one. I expected something elegant like the T-34. One day my friend, Heino Kleine, who was later killed, said 'Do you want to take a look at a Tiger? There's one outside.' I was disappointed. It was a monster – all corners, not at all elegant; the first glimpse was really disappointing.'

(s.Pz.Abt.503) and Michael Wittmann (s.SS-Pz.Abt.101) became publicly known figures as tank 'aces' back in Germany and were highly decorated. Otto Carius, awarded

BELOW Michael Wittmann (left) and his crew in Russia, January 1944. Balthasar Woll, Wittmann's gunner, stands to his left. Both wear the Knight's Cross around their necks. Otto Carius, also a recipient of the Knight's Cross, said wearing it had advantages when requesting supplies or support. *(Ullsteinbild/TopFoto 147823)*

BELOW Otto Carius, a Tiger commander during the Second World War, gets back into the Tiger at the Tank Museum.

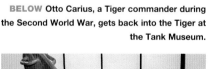

A destroyed Sherman in Normandy, 1944.

ABOVE Two Tigers of the Panzergrenadier Division *Großdeutschland* in Romania in April 1944. The left-hand vehicle clearly shows a neatly applied pattern of *Zimmerit* anti-magnetic paste. (Ullsteinbild/TopFoto 001828)

Leutnant *Alfred Rubbel*,
Schwere Panzer Abteilung 503
'You had to be able to fire instantly. So the gun was always loaded with an armour-piercing round set to a range of 1,000m. That was okay with the flat trajectory and the loader had another 2–3 rounds on the turret floor with him. That was preparation for action.

'The Sherman was not taken seriously, particularly since the Americans gave the Russians about one tank in three. The Panzergranate 43 could go right through both sides so I told my gunner to use a high-explosive round. There was a delay in the fuse of 0.02 seconds, which you could set with a coin. The round then went through the armour and exploded inside and blew their hat off. In other words, it blew the turret off. They were a joke, Sherman tanks.'

Unteroffizier *Doctor Franz-Wilhelm Lochmann*,
Schwere Panzer Abteilung 503
'We [also] didn't listen when the radio said "turn right" – often these were German-speaking Russians. They did that: we saw it all.

'Of course, we also had our tricks. That goes without saying. With an all-NCO crew you could get away with this, but not if there's a boss there. One example: our tank was ordered from one place to another to help the infantry because the Russians had broken through. And while we were there, a large Russian attack with tanks landed right at the place we had just left. Between us and them was a small wooded area. Now, an officer would surely have said "I must get back to where it's all happening" and been off straight away. But we drove past the wood, behind the Russian tanks, alongside the Russian infantry who waved to us. Before they noticed what was happening, we had knocked out 3 or 4 of the 7 or so tanks. One fought back and the rest got away. Of course, it was a stunt but you use your head and do what you can.'

LEFT A Tiger with a whitewash camouflage scheme passes a burning Russian village on the Eastern Front. The front road wheel has been removed, a crew modification to prevent the build-up of earth or snow and the potential loss of a track. (Ullsteinbild/TopFoto 109475)

the Knight's Cross for his success as a Tiger commander on the Eastern Front with *s.Pz. Abt.502*, made the telling point he thought the figures claimed by propagandists were wildly inflated. He also added 'I never claimed one tank – it was my gunner that did the firing – if you had a good gunner you were a success.'

However, the dramatic stands and undoubted bravery shown by such tank men cannot disguise the overall outcome of the operations they were part of. The individual engagement might be won by a Tiger but Allied T-34s or Sherman tanks appeared in greater numbers and would continue the campaign, eventually securing ultimate victory.

Tiger mobility

The Tiger's engine and drive train is – and was – much discussed in military literature since the war, the general consensus being that the tank was too heavy, immobile and suffered from poor reliability. When trying to assess the success of the Tiger, the information that is now available from combat reports can provide a very different picture to that so readily accepted and easy to believe.

Before looking at how the Tiger worked it is important to understand what is meant by tank mobility. Mobility for tanks can be placed into three tiers.

Strategic

At the strategic level there is the challenge of getting the tanks from the barracks, training grounds or factories to the general area of the campaign and this might involve trains, boats (as with Tiger 131 to North Africa) or road-bound tank transporters. In the Second World War, the movement of armour by air was experimented with but was only used in limited circumstances and no plane then existing could have lifted a Tiger.

In Europe, the railway load width system, formalised by the Berne International Load Gauge, allowed a maximum width of 3.15m. This facilitated for the free transportation of

ABOVE A press photograph of Tiger 142 (*Fgr* 250015) shortly after its arrival in Tunis with the 501st Heavy Tank Battalion. The tank is travelling on its transportation tracks and the gun lacks its muzzle brake. Other images taken at the same time show the muzzle brake fitted and covered, so it may have been touched out in this photograph for publication. The image was published in the *National Zeitung* of 11 December 1942 and was being reported on in British Intelligence circles on 5 January 1943. (*Ullsteinbild/ TopFoto 209532*)

LEFT A Tiger of the 504th Heavy Tank Battalion in a Sicilian village. Only one of 17 Tigers escaped from Sicily back to the Italian mainland. (*Ullsteinbild/ TopFoto 048119*)

the load around tunnels and bridges and safe passage past any loads coming in the opposite direction of a similar maximum width. It was possible to travel with wider loads but the considerable planning required and the potential inconvenience to the essential rail networks was considered too intrusive. In consequence tank designers took the 3.15m limit as the maximum width to use in Western Europe (in Britain the loading gauge was narrower, creating a 2.67m maximum load width).

Road transport had a looser set of criteria but generally most European roads could cope with 2.5m-wide loads but the network of roads

was far more extensive to allow alternative routes. Road routes for tank transporters had to be planned anyway as the weight of the Tiger was beyond the loading of many smaller bridges.

Ships and ferries had the potential to carry huge loads but the loading and offloading port facilities for tanks as big as the Tiger needed consideration. There was also the dominance of the seas by Allied Navies for the Germans to take into account, but not one Tiger was lost in the ferrying of vehicles by sea from mainland Europe to North Africa.

Operational

The second tier of mobility is operational i.e. within the area that combat may occur. Approach marches to battlefields and staging areas means the tank may have road or cross-country journeys to complete. Here, the larger and heavier the tank the more fuel is used (especially when travelling across country), the more limited the number of routes that would be available (due to bridge restrictions), or more engineer resources would be required shoring up and preparing the route. The longer a tank drove the more wear occurred and therefore the more likelihood of a breakdown or maintenance issues arising. Therefore it was considered essential to get a heavy tank as

close to the battlefield area by rail or transporter as possible.

Battlefield

Battlefield mobility, where contact with the enemy is actual or likely, can be considered the third tier. In battle all sorts of ground terrain may occur so a vehicle's ground pressure can come into play – how well do the tracks distribute its weight across the ground? Its power-to-weight ratio can also be important here – a tank may use its speed to lessen the chance of an enemy hit as it advances or moves from cover. Alternatively, mobility may be influenced by the level of armour protection. If a tank does not need to fear the threat of enemy weapons on the battlefield, it may take a route forward that others would need to avoid. In this area the Tiger had tremendous mobility when it first appeared in combat.

How mobile was the Tiger?

It is not always a good idea to compare tank with tank in a 'top trumps' manner as different tanks were designed and employed in very different ways. However, the comparison of mobility statistics between the Tiger, the Sherman and the Soviet T-34/85 is salutary and helps re-evaluate the reputation for a supposed lack of mobility the Tiger has gained.

	Tiger	Sherman M4 (mid-production)	T-34/85
Max speed	45kmh	38kmh	55kmh
Average road speed	40kmh	33kmh	47kmh
Average cross country	20–25kmh	17–32kmh	19kmh
Radius on road	195km	193km	260km
Radius cross country	110km	160km	209km
Turning radius	3.44m	18.6m	Skid turns
Trench crossing	2.5m	2.25m	2.5m
Fording	1.6m	1m	1.32m
Step climbing	0.79m	0.60m	0.73m
Gradient climbing	35 degrees	60 degrees	35 degrees
Ground clearance	0.47m	0.42m	0.40cm
Ground pressure	0.735kg/cm²	0.96kg/cm²	0.85kg/cm²
Power-to-weight ratio	12.3hp/ton	12hp/ton	15.6 hp/ton

Reliability

The reliability of the Tiger has also been called into question. The rushed design programme and the lack of time for genuine trialling of the completed vehicle led to inevitable failures and snags. The failure of seals and gaskets was reported in early vehicles, as was the breakdown in the overstressed drive train and new engine. However, Tom Jentz, has shown from German vehicle returns from May 1944 to March 1945 that the operational availability of the Tiger at the front was as good as the Panzer IV. At times it was better than the Panther with an average of

ABOVE Deception. A mock-up Tiger from canvas and wood 'hides' in an Italian orchard. Empty packing cases add verisimilitude.

**Leutnant *Alfred Rubbel,*
*Schwere Panzer Abteilung 503***

'The leap from mechanical to hydraulic transmission and steering was not fully developed. On top of that, the huge final drives were outside the tank and very weakly armoured. So an artillery hit nearby would break something and you could not drive on. These were the problems. The engine was so underpowered it overheated very easily. If you went fast on the road (although at cross-country we really crawled) we would say "let's get a move on" and if we went for 30km [18½ miles] at 50kph [31mph], something would break in the engine. It was the case that if a driver drove without damage for 1,000km [621 miles], he got 14 days leave.'

RIGHT Tigers of the 101st SS Heavy Tank Battalion on the move in Normandy, June 1944.

Unteroffizier *Doctor Franz-Wilhelm Lochmann, Schwere Panzer Abteilung 503*

'A complete company set off with 14 tanks on the way to the Front, then there were 12. After the first day there were 6. Then after 3–4 days there were just 4 or 3 tanks operational. The rest failed because of technical problems. At the beginning that was really tough.'

70 per cent of Tigers being available on the Western Front (compared to 68 per cent of Panzer IVs and 62 per cent of Panthers) and 65 per cent of Tigers being operationally available on the Eastern Front (compared with 71 per cent of Panzer IVs and 65 per cent of Panthers).

Repair and maintenance

The German military had a very different approach to repair of vehicles than the Western Allied Armies. American mass production allowed for a policy of replacing worn parts or vehicles with new, whereas the

RIGHT The replacement of an engine in the field was never an easy task. Here is a Kfz 100 recovery with the standard 3-ton crane. The dog looks a little bored with it all.

SS soldiers load a Tiger with *Pzgr* 39 armour-piercing rounds from a truck parked beside the vehicle. The turret has been reversed, with the gun now over the engine decks. *(Ullsteinbild/TopFoto 041335)*

Leutnant *Alfred Rubbel, Schwere Panzer Abteilung 503*

'We took care to treat our tank like the apple of our eye. Our driver, Eschrich, would really have liked us to take off our boots when we got into the tank so we didn't break anything. We were always ready to help, unlike some other crews who let their driver get on with it alone, we all worked. I came up with the equation: one hour in action made 10 man hours of work – refuelling, re-ammunitioning, track-tensioning – everything you could do in the field. We couldn't do anything like major repairs, but track tension – if that wasn't right when you reversed and turned, then it threw a track, then it was all over.'

LEFT After firing the gun required cleaning. Light oil was poured onto the brush and the five rods screwed together to remove burnt residues from inside the gun. If left, they would erode the barrel, thereby shortening its life and accuracy.

Germans simply did not have the capacity to make parts available in a similar way. When on campaign, especially in North Africa or Russia when hundreds of miles from the homeland, local repair was potentially the only option. Early in the war, military vehicles were returned to civilian workshops for repair (many private garages or small factories related to the motor industry lost all civil trade and could turn to the military for business). Civil contractors also set up workshops in occupied countries.

In the field the general policy was to repair as near to the battle front as possible. A number of repair facilities existed starting with

LEFT The *Fries-Kran* was so named after the makers, J.S. Fries und Sohn. Used by the workshop maintenance crew, a generator was attached to the gantry to power the lifting capacity of 15 tons. Angled metal bars were fixed to the three turret lugs to clear the lifting chains from the turret edges and, when clear, turrets were often rested on extemporised stands made from fuel drums.

RIGHT This turret lift with the use of a *Fries-Kran* shows the turret floor suspended from its three brackets.

the tank crew. At company level Tiger units had a maintenance unit and a larger battalion workshop crew attached to the Headquarters. More serious repairs might mean a vehicle was taken to a central Army Group workshop (often with the driver accompanying the vehicle to assist in the work) or if necessary and when possible, rail transport back to Germany.

The nature of the job would be assessed, sometimes in terms of hours needed for repair and sometimes on the nature of the particular system requiring attention. In the German forces there was a greater tendency to repair, improvise and if necessary actually make components to keep a vehicle running. Commanders were reluctant to relinquish

RIGHT Major repairs are carried out inside an improvised workshop facility. This Tiger's turret is being lifted by an overhead crane – the metal brackets clearing the lifting cables from the turret sides can be seen in use.

ABOVE Removing a broken drive sprocket with the use of a crane on a recovery vehicle.

RIGHT *Zerstörerpatronen*. The left tube, marked (a) in this Allied wartime manual, housed the Z 85 demolition charge, marked (b). Each charge was issued with two igniters held in the cover (g).

DEMOLITION CARTRIDGES TYPE Z
SPRENGPATRONE Z

(a) Steel container for Spr. Patr. Z 85 with air-tight clip-down lid.

(b) Spr. Patr. Z 85. Note hinge on top which holds metal bar for pushing cartridge into gun-barrel.

(c) Funnel-shaped container and lid for Spr. Patr. Z 72. The accessories tin is inserted in wide portion at the top.

(d) Spr. Patr. Z 120.

(e) Spr. Patr. Z 34. This is the type Z cartridge without hinged metal bar.

(f) Ex. Spr. Patr. 34. Wooden training model for (e) above.

(g) Tin with accessories issued with each Z-cartridge. Igniter-detonator set, and one spare igniter and brass connector.

vehicles, as with the distances involved to the homeland replacement was not always an option.

The lack of adequate recovery services led to a high number of Tigers being destroyed by their own crews rather than run the risk of them being captured by the enemy. During the first engagement in Russia in August 1942, three of the four vehicles broke down but all were recovered. During their next combat action, again three of four tanks broke down but one could not be recovered due to its exposed position and only after permission from Hitler was it blown up to prevent potential capture.

From July 1943, crews were issued with destruction charges or *Zerstörerpatronen* to blow up tanks in case they were unrecoverable. They were told 'It is forbidden to allow a repairable Pz.Kpfw.VI "Tiger" to fall into enemy

hands.' The Z 85 charges were carried on the cross-member behind the driver and radio operator's seat. It was suggested one charge be placed in the breech of the gun and one in the engine bay.

Changing tracks

The tracks and running gear caused an estimated 75 per cent of mechanical problems on tanks and the Tiger with its array of road wheels and two types of track was no exception. If a road wheel lost its rubber or had damaged areas of rubber, it could cause vibrations and damage to the swing arms and dampers. The layout of the road wheels, six deep interleaving on each side, led to the possible scenario of 14 road wheels having to be removed to get to one of two wheels of the innermost layer.

ABOVE *'Eins, Zwei, DREI!'* Track is shifted by hand to straighten it ready for replacing on a Tiger in a chilly looking Russian village.

RIGHT A Tiger crew begin to remove the outer road wheels of this heavily draped tank.

Obergefreiter *Siegfried Schiller, Schwere Panzer Abteilung 503*

'One man had to dismount and check the drive train – tracks, wheels. In the case of the Tiger's tracks, the track links were joined together with track pins. When you drove along the pins worked their way out. The track ended here and the pin was out there so they had to be hammered back in again. There was a 25-kilo hammer for this. A man was detailed to do this; each blow knocked in about a centimetre.'

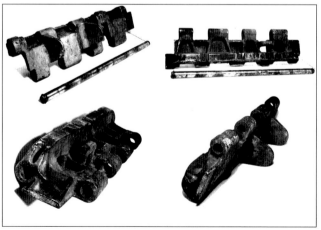

RIGHT Each track link came in two parts with a connecting pin.

LEFT A *Bergepanther* and *Sturmmörser Tiger* at Chobham for analysis after the war. The winch taken from a *Bergepanther* was in use at Chobham until the 1990s. It has now been re-united with a *Bergepanther* in Germany.

To remove the tracks, the track tensioning socket had first to be loosened and then a connecting pin taken out. At 30kg a link, two men could lift three links maximum but a further vehicle might be needed to assist pulling off a broken set of track. Each connecting pin had a narrow hole drilled through the diameter of the pin at the end. Once a securing washer had been placed over the end of the pin to stop it moving, a thin securing pin was hammered into the drilled hole. A drift would be needed to release these pins when the track required breaking. It could take 25 minutes to change battle tracks for transport tracks but up to 36 hours to change inner road wheels. Each wheel was secured with six bolts that would need a large spanner or cross wrench to loosen. Swing arms (two types) were connected to single or double wheels that in turn had differing spacer units between them and the outer layer of wheels.

Once the wheels had been replaced the track could be pulled back on using the 14mm-thick, 15m-long, steel track cable to pull the track forward using the front sprocket as a winch. Tensioning the track to the correct level was possible using the tightening socket, four fingers spacing between the first road wheel and the track was stipulated as correct.

Recovery

It is probable that Tigers got bogged down or stuck no more often than any other German

LEFT The *Zugkraftwagen*, the 18-ton recovery vehicle or prime mover. Three or more of these vehicles might be needed to recover an immobile Tiger. This example, captured in North Africa, prepares to take a jolly bunch for a joy ride.

of unarmoured 18-ton *Zugkraftwagen*, large half-track vehicles that would attach two rigid tow bars, or steel cables from a rear tow hook, to the front two towing shackles of the Tiger. The 32mm-thick, 8.2m-long tow cables were stowed on the top hull surface of the tank but many photographs show the cables being fitted to the front towing hitches and then draped back over the hull so they would be instantly ready for use in an emergency.

It could regularly take two or three or even up to six *Zugkraftwägen* to recover a Tiger and then on descents, an anchor vehicle, such as a Panzer III might be needed to brake and control a Tiger from behind. Each platoon was issued with three recovery *Zugkraftwägen* and the Battalion Headquarters had further resources, but there are consistent complaints in reports on the failures of recovery vehicles to cope and not all units had a full complement of vehicles.

BELOW Tow cables are attached to the vehicle and then draped back over the engine bay ready for swift use in action.

tanks – their problem was a lack of suitable recovery vehicles for so heavy a tank. Only midway through the war did the Germans employ an effective armoured recovery vehicle – the SdKfz 179 *Bergepanther*. Until then the recovery of the Tiger relied on the availability

Living with the Tiger

**Obergefreiter *Siegfried Schiller*,
*Schwere Panzer Abteilung 503***

*'I was the radio man and the radio man was
responsible for the physical well-being of
the crew, which means I made breakfast,
sandwiches, and I cooked everything.*

*'I had a mattress, a narrow one, about this
long and I laid it over the seat. You could put
your feet up then. I slept very well.'*

**Leutnant *Alfred Rubbel*,
*Schwere Panzer Abteilung 503***

*'I slept in the tank. But when there was no
danger from outside, on the rear deck; it was
warm, especially in winter. The gunner, the
driver and the radio operator had elegant leather
seats – they could stretch out as if they were
in the Hotel Adlon [in Berlin]. The gunner and
commander had narrow seats, almost saddles.
The loader bedded down on the turret floor. But
if it was at all possible, if the combat situation
allowed and there was no action nearby, we
slept outside under the tank or between two
trees, so that you were not run over.'*

ABOVE *'Ich hatt' einen Kameraden'*. **Music at the Front, Welikije Luki,
Russia, 1943. Impromptu singing in groups was a feature of wartime
life for both civilians and the military. It took place with a regularity
that is surprising for a modern audience to comprehend.** *(Ullsteinbild/
TopFoto 047616)*

BELOW **A Tiger crewman takes a nap.**

'Of the 1,354 Tiger I tanks made, only six
have survived. So, what happened to the
other 1,348?'
David Willey, Tank Museum Curator

Chapter Nine

Last of the Tigers

David Willey

At the time of writing there are six known complete Tiger I tanks remaining in the world. By complete, this means vehicles where the majority of the components are still present – hull, suspension, turret and gun. Full details of these Tigers can be found in the Appendix on page 158.

OPPOSITE Tiger *Fgst Nr* 251227 is displayed at the Lenino–Snegiri Military Historical Museum in Russia. The gun barrel is not an 88mm.

Aberdeen
Tiger

RIGHT The Aberdeen Tiger (*Fgst Nr* 250031) during trials and recognition photography at the US Army's Aberdeen Proving Ground in Maryland, USA, from where this Tiger got its name.

OPPOSITE TOP The square welded patch covering battle damage can be clearly seen in this view. The repair had been carried out before capture.

OPPOSITE BOTTOM *Fgst Nr* 250031 photographed during its period on display outside the Proving Ground. The holes in the turret side have been plated over.

RIGHT On the test track at Aberdeen Proving Ground during March 1944.

What happened to the other Tigers?

O f the 1,354 Tiger I tanks made, why
do only six remain? From the current
viewpoint the interest in the Tiger tank, its
iconic – almost mythic – status might have
led one to assume more examples of the
vehicle would have been saved. However, one
only has to imagine oneself back in the early
post-war world to begin to see why a Tiger
might not have the appeal it does today to
the modern enthusiast. First, the tank was an

obvious symbol of German military power and
oppression for those countries that had been
conquered and whose lands had been fought
over. The desire to remove symbols of the
Occupation or dark memories that such items
could revive was clear. Also, certain pieces of
military equipment have a useful role in civilian
life and could be used to assist societies in their
reconstruction efforts after the war. But tanks?
Some were used as tugs or the basis for cranes
or heavy plant, but most were consigned for
scrap after useful items such as tools, engines
and the odd wheel had been removed.

A number of vehicles were taken back to
ranges or research establishments for analysis,
or to act as targets for the testing of new
weapons. The Cold War East/West divide led
to a scramble for information from captured
German wartime material. Scientific discoveries,
manufacturing processes and interviews with
German scientists and manufacturers were
collated and published in the UK in a series of
reports under the heading *British Intelligence
Objectives Summary* (*BIOS*) reports. As the
physical material gave up its useful information,
it was then consigned to back yards, transferred
to museum collections or sent to the ranges as
targets or to the smelter for scrap.

As Western Europe returned to normality,
the battlefields were revisited by soldiers and
their families, seeking out graves and scenes

of past exploits. Military and municipal archives show many images of rusting tanks scattered in the countryside. These slowly disappeared to the scrap man's yard or, in a few cases, were identified as memorial pieces.

The French Army and Resistance units did re-equip with some German vehicles during the last year of the war; immediately after the war the 503rd Armour Regiment based at Mourmelon in north-eastern France had 50 Panthers. German vehicles, including armour, were sent to the Middle East, and those countries such as Finland, Spain and Turkey that had bought German tanks during the war continued to use them in service – some up to the 1960s. Some German armour was reused for placing in defensive positions – in Finland and Bulgaria for example. But no Tigers were reused by post-war armies, except for study purposes.

What chance of owning a Tiger?

As with so much militaria, charting the 'value' of material presents a complex challenge. Historic items can be looked at as having much value – for example cultural, technical, symbolic, financial, even personal. Few items could perhaps symbolise Nazi aggression and power in the Second World War as convincingly as the Tiger. Hitler saw the great propaganda value of the tank and used tanks in rallies and imagery to portray a strong new Germany.

In monetary value, complex items of war like tanks or planes have a high price when purchased by the prime commissioning agent or first user. This price can drop considerably when items are sold on to other armies and services at a later date, when their use by the commissioning army has been expended. The lowest financial value point can be after removal from service when – if there is no immediate military or civil use for an item – it can be sold for its scrap or component value, or sometimes much less if the item is complex or hard to cut up. Viewing any item for cultural, historical or financial value is always going to be a subjective affair. One person might see the vehicle as a symbol of a tyrannical regime, another as a

ABOVE The Saumur Tiger (*Fgst Nr* 251114) on transport tracks.

technological achievement, one as an example of what had to be defeated, another as a collector's dream item of great price.

With the interest in the world wars seeming only to grow with the passing of the years, the search for iconic items has led to a business in hunting down surviving artefacts. The development of the 'warbird' historic military aviation phenomenon seems to have prefigured the search for tanks. A small but well-financed band of collectors have seen German armour as *the* desirable type to collect. In a similar manner to the search for airframes, this interest in tanks has led to the scrap yards, battlefields and military collections of Europe being combed over for surviving examples or parts of vehicles for restoration projects.

The collapse of Communism in Eastern Europe led to a period when, in a search for hard currency, many items of Second World War German equipment made their way into markets in the West. This included material where the true ownership was hard to identify, and with the fluid political and loose legal regimes, criminal elements offered items from museums and collections – a situation that was also reflected in many collecting and cultural areas such as fine art and antiques. Negotiating the genuine compared to the false offer, plus the recovery, transportation and export of an item from countries where bribery became endemic, was a hazardous task. A number of people became victims or had close shaves,

ABOVE An early shot
of the Tiger (*Fgst Nr*
251227) displayed at
the Lenino–Snegiri
Military Historical
Museum in Russia.

with con men and organised criminal gangs in their pursuit of historic tanks.

Examples of some of the hazards involved are regularly seen by the Tank Museum. First, there are the e-mailed offers of vehicles for sale – often the name Tiger is mentioned. When a reply is sent asking for images (the polite, but doubting Thomas approach, is essential), often the images that come back are of items patently not a Tiger tank. If a Tiger image is sent, it is sometimes straight out of a book or a photograph of the Tiger in the Lenino–Snegiri Military Historical Museum. When questioning the vehicle illustrated, explanations are sometimes suggested such as 'sorry we sent the wrong image' or 'yes, but we now own this vehicle', or 'this is the type we have for sale'. The Tank Museum has had the pleasure of putting three different individuals in touch with each other, all trying to claim title to and offer for sale, the same vehicle.

Another common approach is the 'need' for funds to recover the vehicle, which is known to be buried at a given point or be hidden in a forest outside Leningrad. The Tank Museum is sometimes approached directly by e-mail with such cases, or when Western businessmen working in the former Soviet Block countries are approached at a hotel bar by seemingly

disinterested individuals asking if they might know anyone in the West who would like to buy a German tank. More than a few businessmen have sensibly approached the museum to see if it would be interested, or to determine what might be a market value for such an item. The museum gives what best advice it can, usually suggesting certain questions. This often ends the process. One businessman had the courtesy to thank the Tank Museum profusely for making sure he avoided 'what I came to see was obviously a mafia con'.

The Internet also appears to have bred a new type of complete chancer – the museum was sent an image of a plastic model Tiger purporting to be the real tank for sale.

Other sources sometimes cited are stocks of material held at Russian film studios. The Russian film industry certainly did make a number of replica Tigers for use in films, including a massive production about the Battle of Kursk – filmed in colour. Black and white versions of the footage are sometimes mistakenly included in documentaries as authentic. However, again no 'real' tanks have emerged but many uniforms and items of personal equipment have found their way from studio sales onto the collector's market.

Then there are the Tiger tanks that have been built for films such as *The Night of the Generals* (1967), *Kelly's Heroes* (1970), and *Saving Private Ryan* (1998), where T-34/85 tanks are converted to look the part. (For war film buffs real Tigers can be seen momentarily on the Arnhem classic *Theirs is the Glory* (1946) and the Welsh Guards' tank exploits in *They Were Not Divided* (1950).)

These are sometimes seen and reported as real by observers, leading to another string of reports that the hopeful might translate into a possibility of a real, previously unknown tank.

Could a vehicle remain in some shed in a remote military camp? In the UK – despite other Tiger tanks having been brought into the country for evaluation with no definite known fate, it is highly unlikely a vehicle could remain undiscovered. But even here, qualifications need to be made. What of the wartime tanks buried as part of a preservative testing exercise on a UK base? The project was later cancelled and the tanks left in-situ. On a number of ranges, tanks were bulldozed into pits and covered. Could a Tiger ever have been treated in such a manner? Probably not, or by now an image, a diary or a memory would have started a search.

As for the rest of Europe a hidden Tiger might be a possibility, although after more than 60 years the chances are diminishing. But without wishing to sound unduly pessimistic, just be warned against the con-artists and false optimism.

There are of course a number of genuine finds, and amazing footage can be seen on websites like YouTube of the recovery of stunningly complete vehicles – but so far no Tiger. While so many purported offers of a Tiger are wilfully false or at best mistaken, it is possible that one day a genuine tank will re-emerge from a bog or forest; therefore one may yet be offered for sale. If the item is privately recovered the vehicle may be broken to raise more money, as has happened with classic sports cars.

Of course, a found tank may not be put up for sale. National collections, such as that of Germany, lack a Tiger I and the country in which a vehicle is found may wish to retain it. The legal, ethical and taste considerations when dealing with the removal of such material

ABOVE A Tiger mock-up based on a T-34/85 chassis, made for the film *Saving Private Ryan*.

from a battlefield, or potential war grave site, may create understandable complexities and exporting military equipment across national borders is a complex and fraught affair. The interest in a 'found' tank would be enormous, but it might not be cash but political influence that secures the vehicle.

Alternatives

Re-enactor groups have recently seen the imaginative conversions of more recent vehicles into impressive Tiger lookalikes – the T-54 gives a good basis with its similar length and width and wide tracks and dished road wheels. Others are building tanks to scale on a structure to fix over a support vehicle to give mobility. Tracks are perhaps the hardest items to replicate, whether intended to work or simply look the part.

Another scenario to bring a Tiger to life is the reconstruction of a vehicle from a number of recovered parts. What might constitute enough parts to allow a vehicle to be claimed a restoration rather than a replica is a difficult question, and is much debated in other collecting circles such as warbirds and classic cars. Large component parts of tanks are discovered and have been traded. Websites

show images of considerable parts of tanks, some in remarkable condition, and others obviously having suffered from 60 years in the ground.

One day, spurred by collectors, re-enactors and film makers, perhaps enough interest will develop to see the reconstruction of vehicles. This has occurred with the remanufacture of replica Messerschmitt Me 262 jets. Time,

however, may be limited – the era of fossil-fuelled vehicles may be coming to an end and the world security situation may impose itself on collectors. Some countries already ban the private ownership of armour, others are tightening their laws.

But we can all still dream of that moment – 'Here, look what I have found in this shed … chap says he doesn't want it.'

Buying a Tiger

During the Second World War, the Japanese Ambassador in Berlin, General Hiroshi Ōshima, visited the Eastern Front where he was shown a Tiger and later he toured the Henschel works at Kassel where the Tiger was made. In May 1943, the Japanese made an approach to the German authorities regarding the purchase of blueprints and an example of a complete Tiger I (they were also interested in the Panther). There had been various discussions between the Germans and Japanese concerning the possible compatibility and advantages of producing similar war material.

After much debate, agreement was eventually reached on the commercial implications of such an arrangement. Japan bought an example of the tank, chassis number 250455, and the vehicle was readied for shipment. Initially, the idea was to break the vehicle into parts and ship it to Japan from Bordeaux onboard a German U-boat. The German authorities invoiced the Japanese for 645,000 Reichsmarks (RM) as the export price. Usually the German government paid 300,000 RM for a vehicle. Some 80 per cent of the 'profit' from the sale to Japan went to the German military, 17½ per cent was divided between Henschel, Krupp and Maybach for their involvement in the development of the vehicle. The commercial rate charged for the vehicle was itemised and copies of the invoice still exist, giving an interesting view as to the commercial pricing of the components of a Tiger in Reichsmarks. (Historical pricing comparison is always difficult, but a German soldier's monthly wage in the Second World War was 110 Reichsmarks.)

Engine	13,000
Hull	54,000
Gear change mechanism	8,300
Tracks	7,000
Turret	26,000
Turret fixtures	20,000
Ammunition	9,000
Optics	2,900
Radios	3,000
x 2 MGs	1,100
Chassis assemblies	12,400
Gun	22,000

As time progressed, the shipment of the Tiger became increasingly problematic and more unlikely, and the tank sat awaiting its fate. In September 1944 it was 'loaned' back by the Japanese for use by German forces.

LEFT Buying a Tiger: Japanese officers admire *Fgst Nr* 250055.

Appendix David Willey

SURVIVING TIGERS

The Aberdeen Tiger

Fgst Nr 250031.
Completed in November 1942.

The tank was captured in Tunisia, having been used by the 501st Heavy Tank Battalion. The unit was later integrated into the 7th Panzer Regiment, 10th Panzer Division. In the process, the tank's number painted on the turret side was changed a number of times. On capture, it was marked '712'. The tank was one of only two in running order left in Tunisia at the end of the campaign. It was listed by Lieutenant Sewell in a report dated 9 June 1943 detailing all Tigers destroyed or intact. The vehicle was taken by the 188th Ordnance Battalion to transport the tank to the USA for analysis at Aberdeen Proving Ground, Maryland.

The tank was disassembled and subjected to analysis, then reassembled to be driven. Testing of the vehicle was stopped in March 1945 and soon after the war it was transferred to the museum at Aberdeen. Here, the vehicle was sectioned, having armour cut off the left side of the turret and hull to reveal the interior.

Despite culls of material in scrap drives and the changing emphasis of the collections, the Tiger remained on display inside the museum building at Aberdeen until 1967 when the need for new space saw the building converted into a new Test and Evaluation Command HQ. The tanks inside were scattered around the extensive Aberdeen site and some joined the rows already outside. Hopes for a new museum building to house this major armour collection were scaled back and in 1973 a new Army Ordnance Museum was opened with most of the tanks still outside. To protect the interior from theft and vandalism, wire mesh was welded across the side of the tank – later this was replaced by welding thin metal sheeting across the gaps. Any vehicle – however impressive and seemingly solid – will suffer with long-term exposure to the elements. Tinwork will rust through, water will penetrate and if conditions permit, freeze and cause frost damage.

This Tiger had a number of re-paints in a variety of different colours and schemes over the years but suffered. Despite sitting on a US Army base, the souvenir hunters were able

RIGHT The Aberdeen Tiger when still on display in the United States.

to remove items and the elements did their worst. By 1989, when an offer of restoration in return for the loan of the tank was made by the German Auto und Technik Museum at Sinsheim, the vehicle was in poor condition.

The loan of the vehicle was agreed but then began a saga that led to the tank residing with a private collector in the United Kingdom. As this book was going to press the tank was in the process of being returned from Europe to the USA.

Tiger 131

Fgst Nr 250122.
Completed in February 1943.
The Tank Museum, Bovington, Dorset, UK.
http://www.tankmuseum.org

This is the tank at Bovington and is the main subject of this book.

The Kubinka Tiger

Fgst Nr 250247.
Completed in August 1943.
On display at Kubinka, the Russian Army Collection outside Moscow.
http://www.tankmuseum.ru

This tank was originally issued as a *Panzerbefehlswagen* or command tank. As a command vehicle, there were a number of adaptations made to accommodate additional radios (such as the Fu 7 for long-range communication) in the turret, which led to the removal of the co-axial MG34 and the removal of 26 rounds of 8.8cm ammunition.

The tank was converted back to a standard *Panzerkampfwagen* in November 1943. It still retains patches of the original *Zimmerit* anti-magnetic coating that was applied to Tigers from August 1943 until production of the tank ceased in August 1944.

ABOVE The Tank Museum's Tiger 131, United Kingdom.

BELOW The Kubinka Tiger, Russia.

6 June 1944. At the time, the unit was based at St Pol, in the Pas de Calais region. They moved to Normandy in early July and took part in the attritional battles around the much contested feature of Hill 112. The tank was abandoned at some point during the retreat from Normandy and is next found in the use of a French Resistance group under a Captain Besnier, who used a number of other captured German armoured vehicles. They took part in the siege of St Nazaire and acted as an occupation force behind the main Allied offensive into Germany. The tank (named *Colmar* by the French) was later used for evaluation purposes before its eventual transfer to the museum at Saumur. The vehicle has transport tracks fitted and shows some evidence of battle damage.

ABOVE The Saumur Tiger, France.

The Saumur Tiger

Fgst Nr 251114.
Completed in May 1944.
Musée des Blindés, 1043, route de Fontevraud, 49400 Saumur, France.
http://www.museedesblindes.fr

The Vimoutier Tiger

Fgst Nr 251113.
Completed in May 1944.
Roadside memorial at Vimoutier in the Orne Département of Normandy, France.

BELOW The Vimoutier Tiger, France.

This Tiger was issued to the 2nd Company of the *Schwere SS-Panzer Abteilung 102* on

This Tiger was probably in service with the *Schwere SS-Panzer Abteilung 102* when on

20 August 1944 it broke down while trying to escape from the contracting Falaise Pocket. As it climbed a hill outside Vimoutier, it joined four other tanks that could not manage the incline and were also abandoned. The crew of the Tiger blew up charges on the engine deck before heading east. The following day, forces from the 2nd Canadian Division advanced up the road. Abandoned vehicles were bulldozed to the side of the road; the Tiger slipped back over a steep embankment.

After the war most tank wrecks were sold as scrap by the French government, but the awkward position of this Tiger meant that apart from the removal of the engine and gearbox, little else had been touched by the mid-1970s. Then attempts at scrapping the vehicle were halted and instead the wreck was recovered and placed back beside the road to act as a memorial to the battles. It now remains as a tourist attraction and is one of the few tanks that actually fought in Normandy, which remains in the area.

The Russian Range Wreck

Fgst Nr 251227.
Lenino–Snegiri Military Historical Museum, Russia.
http://www.russianmuseums.info/M461

This vehicle is in very poor condition, having been used as a range target and has numerous strike marks. The suspension has collapsed and the original gun barrel was crudely cut off at some point in the past. Currently a substitute barrel (not an 8.8cm gun) has been put in place to give the tank a better impression of completeness.

ABOVE The Lenino–Snegiri Military Historical Museum Tiger, Russia.

BIBLIOGRAPHY

Technical
Chamberlain, Doyle and Jentz, *Encyclopaedia of German Tanks of World War Two* (Arms and Armour Press, 1978)
Fletcher, David, *Tiger! The Tiger Tank: A British View* (HMSO, 1986)
Gudgin, Peter, *The Tiger Tanks* (Arms and Armour Press, 1991)
Jentz, Tom and Doyle, Hilary, *Germany's Tiger Tanks Vols 1–2* (Schiffer Publishing, 2000)
Ogorkiewicz, Richard, *The Technology of Tanks* (Jane's Information Group, 1991)

Tiger units and use
Carius, Otto, *Tigers in the Mud* (J.J. Fedorowicz Publishing, 1992)
Gudgin, Peter, *With Churchills to War – 48th Battalion, Royal Tank Regiment, at War 1939–45*
    (Sutton Publishing, 1996)
Jentz, Tom, *Panzer Truppen Vols 1–2* (Schiffer Publishing, 1996)
Schneider, Wolfgang, *Tigers in Combat Vols 1–2* (J.J. Fedorowicz Publishing, 2000)
Wilbeck, Christopher, *Sledgehammers, Strengths and Flaws of Tiger Tank Battalions in World War II*
    (Aberjona Press, 2004)

The Archive and Library of the Tank Museum, Bovington, hold these books and thousands of documents, manuals, reports, images and accounts of tank warfare. Most of the images used in this book come from the Archive. It can be contacted at Librarian@tankmuseum.org.

Index